Practical A

Springer
London
Berlin
Heidelberg
New York
Barcelona
Budapest
Hong Kong
Milan
Paris
Santa Clara
Singapore
Tokyo

Other titles in this series

The Modern Amateur Astronomer
Patrick Moore (Ed.)

The Observational Amateur Astronomer
Patrick Moore (Ed.)

Telescopes and Techniques: An Introduction to Practical Astronomy
C.R. Kitchin

Amateur and Professional Designs and Constructions

Patrick Moore (Ed.)

The cover shows the glass-fibre-domed observatory built by Brian Manning (Chapter 12)

ISBN 3-540-19913-6 Springer-Verlag Berlin Heidelberg New York

British Library Cataloguing in Publication Data
Small astronomical observatories: amateur and professional
designs and constructions. – (Practical astronomy)
1.Astronomical observatories 2.Astronomical observatories – Design and construction
I.Moore, Patrick, 1923–
522.1
ISBN 3540199136

Library of Congress Cataloging-in-Publication Data
Small astronomical observatories: amateur and professional designs and constructions / Patrick Moore (ed.)
p. cm. – (Practical astronomy)
ISBN 3-540-19913-6 (pbk. : alk. paper)
1. –Astronomical observatories–Design and construction--Amateurs' manuals. 2. Astronomical observatories–Great Britain--Amateurs' manuals. I. Moore, Patrick. II. Series.
QB82.G7S63 1996 96-13224
522′.1–dc20 CIP

Printed in Great Britain

Typeset by T&A Typesetting Services, Rochdale, England
Printed at the Alden Press, Osney Mead, Oxford
34/3830-543210 Printed on acid-free paper

Contents

vii

Introduction

Astronomy is still one of the very few sciences in which the amateur can play a valuable role.

Indeed, amateur work is warmly welcomed by professional astronomers. During the past few decades the whole situation has changed; whereas the average amateur used to own a modest telescope and concentrate only upon various well-defined branches of observation (notably Solar System researches, and variable star work), the modern amateur can make use of affordable but highly sophisticated equipment.

Obviously, the serious amateur will need an observatory, and while there are many books dealing with telescope construction and use there are very few dealing with actual observatories.

The present book will, I hope, fill this gap in the literature. The observatories described here are of various types, ranging from simple run-off sheds to complicated domes; there are observatories designed for studying the Sun, others suited to "deep-sky" enthusiasts, others built for the benefit of radio astronomers or astro-photographers. In each case useful hints are given, and it is hoped that the would-be observatory builder will find a great deal here to help in the construction.

No two observatories are the same; each has its own advantages – and its own drawbacks!

No attempt has been made at "standardisation" of style; each author has been free to write in his own way, and to explain the procedure followed and the various difficulties encountered. Measurements are given in both Imperial and Metric units, with author's own preference coming first.

If you intend to build an observatory – good luck!

Patrick Moore

Chapter 1

A Practical Roll-off Roof Observatory in Michigan, USA

Dennis Allen

My family owns property up in west-central Michigan. This is an area known for its relatively dark skies. It's a place I go to hunt, fish, and enjoy the occasional clear night. Early this spring, I was treated to a whole flock of clear nights. One problem: too much snow on the ground. There was simply no place to set up my telescope.

So this year, I vowed to build an observatory.

My original idea was to create a peaked roll-off roof. This building would have a 12 ft (3.6 m) square

Figure 1.1 Dennis Allen's roll-off roof observatory.

wood floor and 4 ft (1.2 m) walls. Wide enough to leave plenty of room for my 13.1 in (333 mm) reflector. Whenever I got a bigger telescope, something requiring more stability, I could always pour a small concrete pad. I wanted something simple, practical, and durable. But I didn't want to spend years planning and months building.

I kept my design simple: a one-piece roof, rolling to the north. Three inch (75 mm) caster wheels would extend down from each truss and would ride on aluminum channel. To keep the roof light, I'd use corrugated sheet metal. The south wall would have a standard 3 ft × 7 ft (910 mm × 2130 mm) door, cut off at the 4 ft (1220 mm) mark. The upper 3 ft (910 mm) section would hang from the southern gable.

Step one was to build a scale model. Most people do not know what a roll-off looks like. A one-inch-to-the-foot (1:12) scale model helps illustrate your intentions (see Figure 1.2). You can obtain materials to make the model from any model airplane shop.

As it happens, my father is a carpenter. I told him my plans and showed him my model. I kinda knew he'd help! He quickly drew up a list of materials. To keep snow off the roof, he suggested a 6/12 pitch roof. To maintain head clearance, he suggested using church trusses. With 12 ft (3.7 m) church trusses, the bottom 2 in × 4 in (50 mm × 100 mm) doesn't go straight across. Instead, two 6 ft (1.8 m) horizontal 2 in × 4 in (50 mm × 100 mm) pieces connect to a vertical 2 ft 6 in (0.8 m) 2 in × 4 in (50 mm × 100 mm), creating an interior 3/12 pitch.

Figure 1.2 Scale model of the prospective observatory.

As soon as the snow melted, I contracted a bulldozer to clear and level the top of my hill. My dad ordered the trusses, custom made, from the local lumber company. One regular 12 ft (3.7 m) truss (for the northern gable), and three of the 12 ft (3.7 m) church trusses. Meanwhile, I ordered four 16 ft (4.9 m) sections of $1\frac{3}{4}$ in × $\frac{3}{4}$ in (44 mm × 19 mm) aluminum channel from a local sheet metal shop.

By the time I was ready to build, several people told me a small concrete truck could make it up the hill. I always wanted a concrete floor. Concrete makes for a solid foundation, and is less expensive than treated wood. With a concrete floor, my building could house a bigger telescope. To house an 8 ft (2.5 m) long telescope, for example, I'd simply locate its base a few feet north of center. I had considered the thermal problem of concrete. But this is a roll-off, after all. Once opened, the heat should dissipate quickly.

There was one drawback, however. A concrete floor meant a permanent structure. Such a structure would require a special building permit from the local township board. I would have to hire a surveyor to obtain the exact location of the structure. Finally, I would be required to withdraw that location from the Commercial Forest Act of Michigan.

While acquiring the permits, I decided to upgrade my design. I opted for a 12 ft × 14 ft (3.6m × 4.3 m) building with 5 ft (1.5 m) walls. I would have liked a 14 ft (4.3 m) square building, but I already had the 12 ft trusses. These trusses were designed for 4 ft (1.2 m) centers. So my dad made a fifth truss, using the other trusses as a pattern, to give me 3 ft 4 in (1 m) centers.

By the time I got my permits it was almost the end of June. But with help from my dad and brothers, I knew it wouldn't take long to build. In fact, it didn't take an hour and we already had the forms in the ground. Once the forms were down, I had the local cement company bring in three cubic yards of concrete. We went with a 4 in (100 mm) thick floor, 10 in (300 mm) edges. We used 5-gallon (23-litre) buckets, open at both ends, as forms for the outside rail posts. The whole process took only a half day. There was plenty of leftover concrete, though no extra forms. We should have poured an outside viewing pad, something you might want to keep in mind if you decide to pour concrete.

Actual construction started a couple of days later. On the first day of construction, we threw up the walls and the rails (see Figure 1.3). The walls were built out of simple 2 in × 4 in (50 mm × 100 mm), 16 in (410 mm) centers. We used treated pieces of 4 in × 4 in for the top of the walls, the bottom of the trusses, and the outside rails. To connect the rails to the walls, each piece of 4 in × 4 in (100 mm × 100 mm) had a 2 in (50 mm) square notch at the end.

On the second day of construction, we put up the plywood. Originally I thought of using cheap particle wood (chipboard), covered with vinyl siding. My dad, however, talked me into using fake rough-cut 7/16 in (10 mm) plywood. This material looks like rough-cut pieces of 2 in × 8 in (50 mm × 200 mm). As it turns out, this material is stronger than particle wood and already had a gray primer coat.

We brought 13 sheets of plywood. The sheets were cut with a 2 in (50 mm) overhang on the bottom and a 6 in (150 mm) overhang on top. The top overhang turned out to be a blessing. It would end up overlapping the 4 in × 4 in (100 mm × 100 mm) roof beams, covering the caster wheels completely, thus keeping the elements out. As a bonus, this top overhang would serve to keep the roof rolling in a straight line.

We brought a full-size door, cut at the 5 ft (1.5 m) mark. So to finish the day, we hung the bottom section. We made this section of door swing to the outside, thus preventing people from kicking it down. If you hang a door this way, however, remember to use special outdoor hinges.

Figure 1.3 Construction of the walls and rails.

On the third day of construction, the roof went up. We mounted the ten 3 in (75 mm) caster wheels on the two 14 ft (3.6 m) pieces of 4 in × 4 in (100 mm × 100 mm). The caster wheels were spaced so that each wheel would rest under a truss. The channel was used to make sure the wheels were lined up correctly. This channel was already counter-tapped, so we quickly screwed it onto the rails.

One suggestion: keep your location in mind. Apart from a portable generator, we had no electricity. So try to have as much of your material prepared off-site as possible.

The two 14 ft (3.6 m) pieces of 4 in × 4 in (100 mm × 100 mm) were dropped into each channel and the trusses placed on top. We used 14 ft (3.6 m) pieces of 2 in × 4 in (50 mm × 100 mm) to connect the trusses. After some adjustments to the trusses and caster wheels, we could roll the roof back and forth.

Originally, we ordered 16 ft (4.9 m) pieces of 2 in × 4 in (50 mḿ 100 mm) to mount the corrugated sheet metal. We didn't stop, however, just because we were stuck with 14 ft (3.6 m) pieces. To get our north-south overhangs, we simply used scrap pieces of 2 in × 4 in (50 mm × 100 mm). This added a little weight to the roof but hey, if you stop construction for every minor inconvenience, you'll never get any work done, will you?

For the roof, we used eleven panels of 8 ft (2.4 m) White McElroy. These sheet metal panels went up in only a couple hours. We did have to cut one end-piece. For that, however, a roofing knife did the trick. Simply run a straight edge with the knife and flex the sections until they split. But whatever you do, be careful: all the panels have a smooth edge, but the cut pieces are razor sharp!

Here's another useful tip. When you install your panels, do both sides at the same time. Each time you have enough panels, put a section of cap on. When we installed our panels, we left the cap to last, which wasn't easy. Being the lightest in weight, I had to perform a high-wire act just to get the caps nailed down.

On the fourth day of construction, we worked on the gables. We were running short of plywood, and had to buy three more sheets. Which, as it turned out, was about how many sheets worth of scrap we had left over!

The northern gable was easy. The fake rough-cut plywood was measured and cut to butt right up to the corrugated sheet metal. It was notched out for the 2 in × 4 in (50 mm × 100 mm) slacks. To keep out the elements, we left a few inches of overhang on the bottom of the gable.

The southern gable was a different story. I wanted 3 ft × 7 ft (910 mm × 2130 mm) of clearance for the door. To achieve that, we couldn't place a piece of 4 in × 4 in (100 mm × 100 mm) across the threshold. The fake rough-cut strengthened the walls considerably, but the southern wall was still the weakest. So for more strength, I decided to add tables to each corner on the southern wall (see Figure 1.4a).

For the upper section of door, we built a 2 in × 4 in (50 mm × 100 mm) frame. For strength, we used a couple of 2 in × 4 in (50 mm × 100 mm) struts to connect the lower gable corners to the next adjoining truss. We placed our hinges at the top of the upper door, so that it would swing inward. When I want to move the roof, I simply prop the upper door with an extra piece of plywood (see Figure 1.4b). To lock the upper door, I mounted I-bolts and drilled two holes into the 2 in × 4 in (50 mm × 100 mm) frame.

To roll the roof off, there couldn't be any plywood overhang on the southern gable. So we used 1 in × 6 in (25 mm × 150 mm) trim, nailed to the southern wall, to cover the crack. We also used this material around each section of door (see Figure 1.5).

To keep the roof from blowing off, I installed chain binders to each corner of the building. These chain

a

b

Figure 1.4 The interior of the southern wall.
a Tables added to the corners for strength. Chain binder also visible, *left*.
b The upper section of the door propped open.

Figure 1.5 1 in × 6 in trim added to the southern wall and door frame.

binders hook to big eye-screws, which are screwed into the roof's 4 in × 4 in (100 mm × 100 mm) pieces.

And that's it! Since then, most of the work has been minor. For security, I installed a latch guard on the bottom door and a 12 ft (3.7 m) cattle gate at the bottom of the hill. They may not stop anybody from breaking in, but they should make people think twice.

I added 40 in (1 m) strips of 4 in (100 mm) square foam between the trusses and the pieces of 4 in × 4 in (100 mm × 100 mm). They keep the elements out, as well as animals and insects. This last month, we've had lots of rain in Michigan. The building, however, has remained bone-dry.

Figure 1.6 Side view of the observatory and external rails.

Figure 1.7 External rails with the roof partly rolled.

As an added touch, I installed a 12 ft × 14 ft (3.7m × 4.3 m) piece of outdoor carpeting. The carpet helps protect your telescope from the corrosive effects of concrete, and saves that occasionally dropped eyepiece!

Were there mistakes? Most certainly. When the cement truck left, he had to dump the extra concrete. As I said, that concrete could have been used for another viewing pad.

We could have reduced the weight of the roof if we had single 16 ft (4.9 m) strips of 2 in × 4 in (50 mm × 100 mm). In fact, we could probably have gotten away with 16 ft (4.9 m) 2 in × 2 in (50 mm × 50 mm) strips (although the structure has to be within the local building code).

If I had to do it over, I'd have used 4 in (100 mm) caster wheels instead of 3 in (75 mm) wheels. The 3 in (75 mm) wheels have already developed a fine film of rubber, probably due to wear and tear; and at some point I may end up replacing them.

But there were pleasant surprises. The plywood overhangs cover the caster wheels rather well, and made building the roof easier. In addition, I don't have to insert foam strips between the caster wheels to keep the weather out.

The church trusses make the inside look like a cathedral (see Figure 1.8). Had I known I'd have that much head room, I'd have stuck with 4 ft (1.2 m) walls.

Figure 1.8 The church trusses supporting the roof.

I was a little worried about the channel. The caster wheels are $1\frac{1}{2}$ in (38 mm) wide, while the channel is less than $1\frac{3}{4}$ in (44 mm) wide at the ID. I figured for sure the wheels were going to bind. As it turns out, however, the tight channel keeps the roof running in a straight line (Figure 1.9), and there is no need for side casters.

At first the roof was very hard to roll. I was already thinking I might have to rig up a block-and-tackle system, but as time went on, the rolling became easier. The plywood overhang tends to swell, so I've been inserting wooden shims to keep it peeled back. Applying silicone spray to the caster wheels also helps reduce friction.

Figure 1.9 The observatory with the roof fully rolled back.

Conclusion

The entire building cost about $1500 in materials, which was less than I expected. The success of this project goes in large part to having a carpenter supervise the construction. I'm very lucky to have one for a father! I'm also lucky to have brothers willing to lend a hand.

If you don't have a relative in construction you should consider hiring one (a construction worker, not a relative). You'll cut down on the building time and you'll end up with a better observatory. You know the old saying, "pennywise and pound foolish": if you need to save money, get your friends and family to help with the grunt work.

The Future

At some point in the future, I'll probably replace those 3 in (75 mm) caster wheels with 4 in (100 mm) ones. But new wheels call for new channel and for the moment, I'll just keep the wheels cleaned and greased.

Except for your head, the 5 ft (1.5 m) walls provide a good protection against the wind. In the future, I think I'll make a couple of 3 ft × 3 ft (900 mm × 900 mm) wind panels. These panels will have 2 in × 4 in (50 mm × 100 mm) pegs about 2 ft (600 mm) long. They should work like side rails you put on a truck bed. In whatever direction the wind blows, I'd just put up panels to block it.

Building a Larger Observatory

If you need to build something bigger, I wouldn't recommend building a two-piece roll-off. In a two-piece, your rails run east and west. Both east-west gables would have to overlap the walls. You'd have no choice but to install the door on either north or south wall. Since you need a solid 4 in × 4 in (100 mm × 100 mm) for the channel, you'll have to have 5 ft (1.5 m) walls to install a 5 ft (1.5 m) door.

Another disadvantage of a big building made to this design would be the elements. To close and open the building, you would need to insert and remove strips of foam where the two roofs overlap. Otherwise, you invite rain, snow, birds, insects, and other assorted critters.

No, stick with the one-piece design. Simple, yet weather tight. If the roof is too heavy to move by hand, rig a block-and-tackle system. You could even try installing an electric winch or a garage door opener.

Recent Events

It's been ten months since we built the observatory and thus far the building appears in good shape. The inside stayed dry all winter. The outside rails, however, did need some work. The 4 in × 4 in (100 mm × 100 mm) wood was a little green and the west rail twisted. I had to shim the center post and add a few reinforcing trusses (something I should have done in the first place).

I did notice one other problem. Since last fall, the roof was getting harder and harder to roll. Straightening out the outside rails helped, but then I noticed the distance between the east and west channel wasn't built even: the mid-section of the building loses about $\frac{3}{4}$ in (19 mm). I also noticed the rollers on the east side appeared to be staggered against the lips of the aluminum channel. The rollers are $1\frac{1}{2}$ in (38 mm) wide, while the channel is only $1\frac{3}{4}$ in (44 mm) ID, which leaves little room for error. So I decided to replace the east side with 3 in (75 mm) wide aluminum channel. It worked! Rolling is much easier now. I imagine replacing the west side with 3 in (75 mm) channel would make rolling extremely easy – but then I'd have to add side rails to keep the roof rolling in a straight line. A lot of fuss that, in my opinion, wouldn't be worth the effort.

The Telescope

Currently, my observatory houses a 13.1-in (333-mm) f/4.5 Coulter Odyssey Dobsonian (see Figure 1.10).

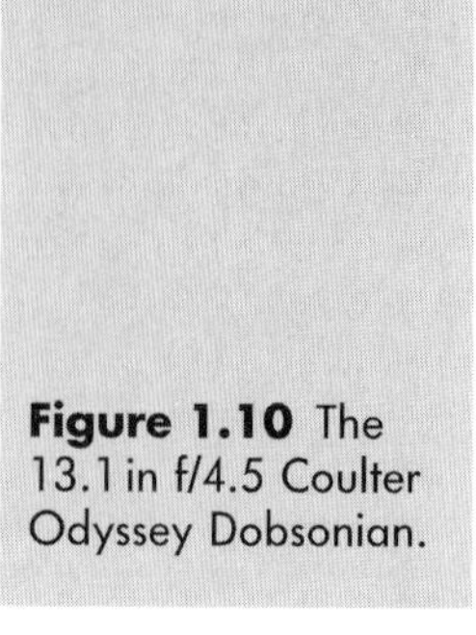

Figure 1.10 The 13.1 in f/4.5 Coulter Odyssey Dobsonian.

I've modified this telescope somewhat. Installed a Novak mirror cell and diagonal holder, and also an AstroSystems phase-two focuser. I had the primary mirror checked and refigured by Galaxy optics and replaced the secondary mirror completely. I added an 8×50 finder, a telerad, and an NGC-SKY MAX computer. In addition, this telescope sits on top of an equatorial platform made by Tom Osypowski. All in all, quite an enjoyable unit.

Recent events, however, dictate change. In June, I got hold of a retired optician through CompuServe. He had a 2 in (50 mm) thick piece of Cervit he was willing to grind. After much discussion, I had him start work on a 24 in (600 mm) f/4.5. I also ordered a 24 in truss tube kit from AstroSystems. This telescope will be about 8 ft (2.4 m) long. A little tight, but I think it will fit in my observatory.

In November, I finished building the mirror and rocker boxes. The 24 in mirror was completed in December and shipped to the coaters, and I have since received it back. Final assembly will begin as soon as the weather breaks. In the future, I plan to order another equatorial platform. Eventually, I want to use the 24 in telescope for prime focus astro-photography.

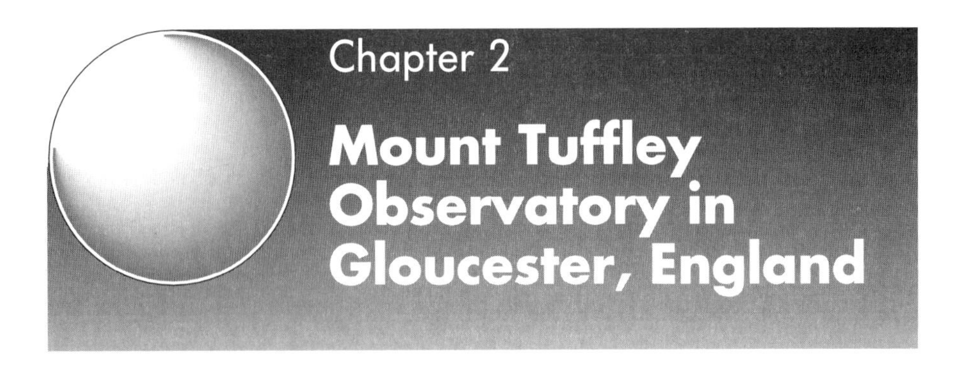

Chapter 2

Mount Tuffley Observatory in Gloucester, England

John Fletcher

Figure 2.1 Mount Tuffley Observatory. Note the corners, giving extra internal space.

Having spent many years on visual astronomy, I became very interested in astro-photography, particularly of deep-sky objects. I soon discovered that astro-photography had a great advantage over visual observation for deep-sky work: one can indeed easily record far greater structural detail in extended objects, and record much fainter stellar objects than can be seen visually. This is the case even when making the shortest of time exposures and using the small-aperture camera lens or telescope.

I had decided that recording deep sky objects was to be my main interest in astronomy, with the possibility of a scientific contribution or that ultimate, a supernova discovery.

My main interest in the photographic search for supernovae required a permanently-housed driven equatorially-mounted telescope that was polar-aligned to perfection. As most supernovae peak at around 16th magnitude it is necessary to carry out either prime-focus or Newtonian-focus photography using the main instrument rather than shorter focal length camera lenses.

I built *three* observatories for my first permanently housed telescope, which was to be a 216 mm (8.5 in) Newtonian reflector. The first two buildings were similar. Each consisted of the lower half of a shed and had run-off roofs, the design of the roofs differing somewhat. These structures were an immense improvement on a temporary site, but, when in use, everything above about 1.2 m (4 ft) was exposed to the elements.

In 1981 I built the third observatory, which is still in use today by one of Britain's foremost astro-photographers, Bernard Abrams. It was an all-wooden octagonal rotating observatory which was later to be named "Mount Tuffley Observatory" by members of my local astronomical society, the Cotswold Astronomical Society.

In 1985 I started thinking of building a larger observatory designed to house a much larger telescope.

I asked my cousin (Mr. Christopher Smith), who is an engineer, if he could design an observatory for me of metal construction. He thought about it for a while and his only question was, "Would you like a proper dome?"

After much planning he started work on it.

His first job was to cold roll into a 3.045 m (10 ft) circle a length of 40 mm × 40 mm × 6 mm (1.6 in × 1.6 in × 0.25 in) black "angle" steel (Specification EN125, BS4360). This was to be the circular rail that the dome was to turn on. To this circular rail were fitted eight evenly spaced thrust and roller bearings.

Having completed the rail he then made another metal circle made from the same steel gauge but with a larger diameter of 3.147 m (10 ft 4 in). This was to become the circular bottom section of an all-metal rib structure which would be the skeleton of the dome. It was made larger so that the bearings would be enclosed, and the outer section of the finished dome would hang over the rail so preventing any rain from entering around the entire lower circumference.

From this point he produced two half circles of smaller diameter than the bottom section, using 30 mm × 30 mm × 30 mm (1.2 in × 1.2 in × 1.2 in) black steel T-section. With the bottom of the dome's circle set horizontal, he welded the two half-circles of the T-section in vertical positions at a distance of 1 m (3 ft 3 in) apart onto the bottom circle. Between these was to be the opening for the observatory shutter, and

Figure 2.2
a Mount Tuffley observatory with its low trap removed.
b The Meade LX200 + CCD and the open shutter.

they also formed the two rails for the shutter to run on. Next, 30 mm × 5 mm (1.2 in × 0.2 in) black flat steel was welded from one side to the other across these two large vertical half circles and at a point offset from the highest position by 370 mm (15 in). This offset was to allow for the viewing of the zenith without obstruction. After this twelve circular ribs were welded around the bottom section and onto the shutter rails at a distance of just under half a metre (18 in) apart. Finally at this point the skeleton of the sliding shutter fitted with 20 mm (0.8 in) diameter bearings was made, completing the skeleton structure.

We then had to work out how the sliding dome shutter could open 370 mm (15 in) past the zenith without overshooting and hitting the building that the dome was to sit on. To allow for this the actual shutter was made shorter than the slit, and a separate lower detachable trap just under a metre (3 ft) square was made. The trap – which is in the lowest frontal position of the shutter when the dome is closed – can be independently removed before sliding the main shutter open. This is very useful when viewing or photographing low altitude objects (Figure 2.1).

Construction

How did it go? Here is my personal account of my own time and the work involved in building Mount Tuffley Observatory. . .

My cousin started work on the design at the beginning of January 1986, and a few weeks later I had the skeleton of the dome delivered to my home on a large flat trailer.

The planning and work was in my hands after this. My first task was to lay a 3.65 metre (12 ft) square concrete base, about 370 mm (15 in) deep. I mixed the concrete by hand myself to save on costs.

The skeleton was then rested on some borrowed milk crates in my garden, to lift it off the ground and make it easier to work on. I applied several coats of red oxide primer to the metal structure to protect it, before covering the skeleton with sheet metal.

I then bought 14 sheets of flat 20-gauge aluminium alloy, 2000 rivets, a hand rivet gun, and four small G-clamps to use as extra hands for holding the sheet metal in place. I was about to embark on the most soul-destroying and tedious job that I have ever done in my entire life. . .

In February I started drilling out some 1700 holes through the 6 mm ($\frac{1}{4}$ in) thick metal rib structure. For several hours each day, for nearly two weeks, in appalling rain and winter weather, I drilled each of the holes some 37.5 mm ($1\frac{1}{2}$ in) apart. Then I had the cutting and shaping of the sheet alloy to do, a job which turned out quite well considering I had no previous experience of metalwork. This was followed by applying a flexible sealer between the sheets and the metal ribs to seal the joints.

I used over 1700 rivets in all and broke or wore out about a dozen 2 mm (0.8 in) drill bits.

By the end of March the dome structure was completed. I then spent time putting aluminium etching primer and five coats of matt grey enamel coach paint onto the dome. I must say it looked very impressive standing there, all gleaming and brand new and complete. I deserved a rewarding sight after all that riveting.

Next came the observatory building. I built a 3.65 metre (12 ft) square structure with corner timbers on top for the dome rail to sit on. The structure was made of 100 mm × 100 mm (4 in × 4 in) timber. The sides consisted of vertical timbers only 300 mm (12 in) apart all round for strength, and added security against possible forced entry through the sides. It stood just over 1.2 m (4 ft) high, with a 1.2 m × 0.75 m (4 ft × 2 ft 6 in) wide oak door at the front. I weatherproofed the outside of the framework with 20 mm ($\frac{3}{4}$ in) tongued-and-grooved timber.

The square structure had considerable advantages over a traditional circular observatory building, such as most observatory domes sit on, in that there is lots of room under the four corners, beyond the diameter of the actual dome circumference, and this space can be used for storage (Figure 2.1).

The final job, prior to lifting the dome on top, was to seal the concrete floor to prevent rising damp. This was done by using builders' house-foundation plastic waterproof sheeting and putting another 100 mm (4 in) layer of concrete on top.

With the help of no fewer than eight people, the dome was lifted into place.

I put in a raised wooden floor which has a fitted carpet to keep the feet warm. This floor wasn't extended to cover the central area. A 1.2 m (4 ft 9 in) area was left bare at the centre so that no floor vibration was transferred to the telescope mounting.

Finally, after more than three hundred hours' labour, I lined the inside of the lower structure with 6 mm ($\frac{1}{2}$ in) marine (resin-bonded) plywood to give the building a professional appearance and help prevent any dampness entering through the sides. The entire inside of the all metal dome was painted matt black to prevent any reflections.

On 21 November 1988 I was honoured by having my observatory officially inaugurated by TV presenter and astronomer Heather Couper – the opening was shown on Breakfast Time BBC TV and Central ITV News.

The Telescope

Until recently a 254 mm (10 in) f/6.3 Newtonian reflector was installed, and was used continuously throughout the last nine years. Indeed, I often recorded 18th magnitude stars, helped by having an accurately polar-aligned telescope permanently housed inside this beautiful dome.

This year, after a long hard struggle to save enough cash to buy a suitable telescope for CCD imaging work, I updated my equipment. Having seen Bernard Abrams' brand new Meade system, and after a little friendly persuasion to part with all my savings, I went ahead and purchased a fully computerised Meade LX200 Schmidt Cassegrain 254 mm (10 in) f/10 reflector.

With this system, observational, photographic and CCD astronomy suddenly becomes very exciting. Locating objects is very fast, so you can cover more of the sky. As before, the dome has great advantages (as does any observatory), for even with this type of telescope, precision polar alignment is better made permanent. The Meade has another, and unexpected, advantage over the somewhat more bulky equatorially mounted Newtonian reflectors: it leaves more room for comfortable chairs and cabinets! (Figure 2.3).

There are, however, some very important things to think of when installing a PC (or indeed any electronics systems) into the damp and often freezing conditions of an amateur observatory.

Figure 2.3 The Meade LX200 Schmidt-Cassegrain.

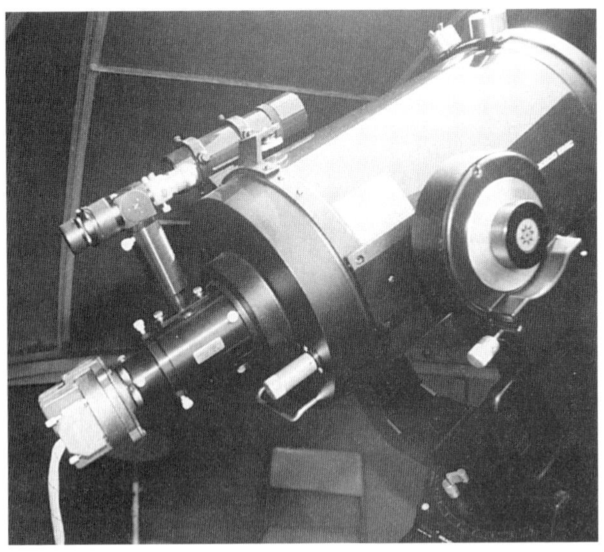

I decided to build an enclosed computer console within the Mount Tuffley Dome. It developed into a large, all-wooden airtight cabinet which is basically divided into 4 sections (Figures 2.4 and 2.5).

In compartments "1" and "2", the 386DX computer and its monitor respectively are totally enclosed with exception of two large vents for warm-air

Figure 2.4 The CCD and computer cabinet under one of the observatory's corners.

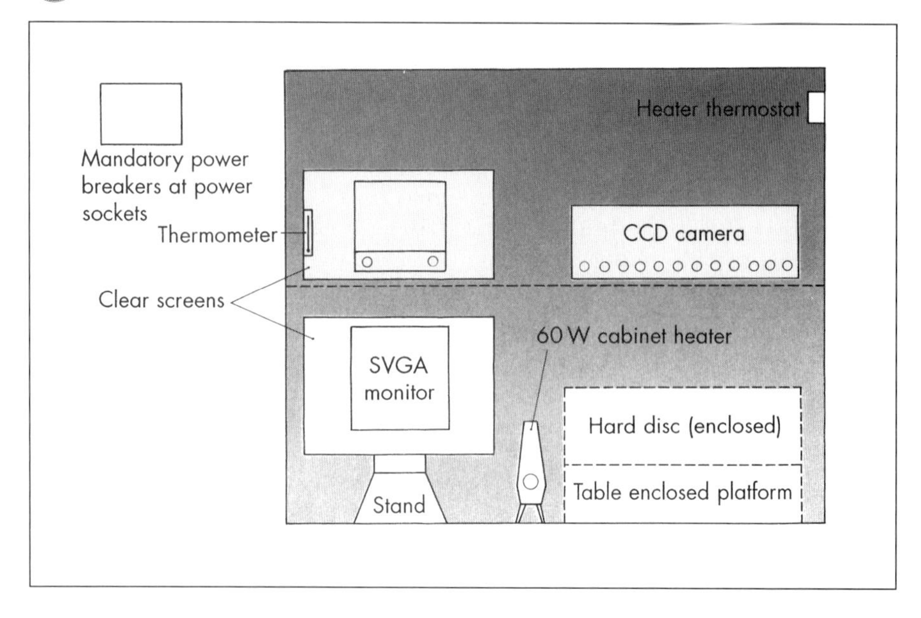

Figure 2.5 Computer and CCD console.

circulation when the telescope is in use and after closing down for the night. In the other upper compartments, "3" and "4", are the Starlight Xpress CCD system, and the CCD TV monitor. These also are fully protected from the elements when I close the observatory down at the end of an observing run.

A small hinged door can be opened for access to the various switches on the CCD imager when it is in use, so there is just a little exposure to the outside elements involved at this point. The cabinet just fits underneath one of the corners of the observatory.

The CCD TV monitor and computer monitor can be viewed through 6 mm ($\frac{1}{4}$ in) clear UV blocking Perspex (Plexiglas) squares set in the front panels. These can be screened by small curtains during exposures.

Finally I should warn of the damage that can be done to a PC if it is exposed to the night elements and kept outside for any length of time unprotected. Cold, and more particularly water condensation, can cause internal short-circuits and at worst the complete loss of the machine.

I have fitted a small electric cabinet heater inside the console. It is only 60 watts output, but controlled by a standard thermostat (for central heating) it can maintain any temperature from 5°C to 32°C (41°F to 90°F). I also have a thermometer fitted. It can be viewed through one of the clear plastic screens: I prefer to keep the temperature around 5°C to 10°C (41°F to 50°F). maximum. Although the heat is almost totally contained I don't want excessive heat rising and escaping into the night air around the inside of the dome, for obvious reasons.

To keep damp at bay, I leave several desiccant moisture-absorbing packs at various positions inside the cabinet.

Observatory Instrumentation

Until you have used one, you cannot appreciate how professionally thought out and how well-made today's 'state-of-the-art' telescopes can be.

An automated supernova search is no problem at all.

The telescope can also be trained to follow a comet's direction for accurate long exposure CCD or photographic work. The beauty of the system is that objects are found in seconds, almost without effort and – as I mentioned – the amount of sky you can cover, even in the shortest of sessions, is incredible. The first night that I used the telescope visually with an eyepiece fitted that gave a 15′ (minutes of arc) field I located forty-one objects in one hour.

These telescopes will, I am sure, turn out regular discoveries and some real scientific work if used to their maximum. By serious work I mean the kind of thing for which professional establishments seldom have telescope time. And with the ability to carry out photometry using the CCD systems, there is yet another field to enter.

But it's still supernova patrol work for me.

I have already seen a 17th magnitude star on the monitor in an exposure of a few minutes using the CCD within a 13th magnitude 10′ × 5′ galaxy. This is almost incredible, and the knowledge that you can move ahead to the next one after checking the field makes it extremely quick compared with photography. Many a night in the past I have had to develop a film at 4.30 am, and then check ten galaxies for supernovae. Ten galaxies means twenty negatives, and using my earlier 254 mm (10 in) f/6.3 reflector, each exposure required a duration of say, five minutes to reach magnitude 16.5. Two exposures had to be made on each galaxy to eliminate possible flaws and dust specks: supernova false alarms waste everyone's time and can cause much embarrassment.

At the end of a long night your concentration has long gone, anyway.

To summarise, an observatory such as I have described, in combination with good modern instrumentation, is for me the ultimate tool for any amateur astronomer. An observatory of this nature gives the observer maximum protection from the elements (apart from having no heating) and so allows for greater comfort, speed of sky coverage, and in many orientations acts as a shield from any of the stray light emitters in your neighbourhood.

Finally, I would like to wish to you all good observing and clear skies.

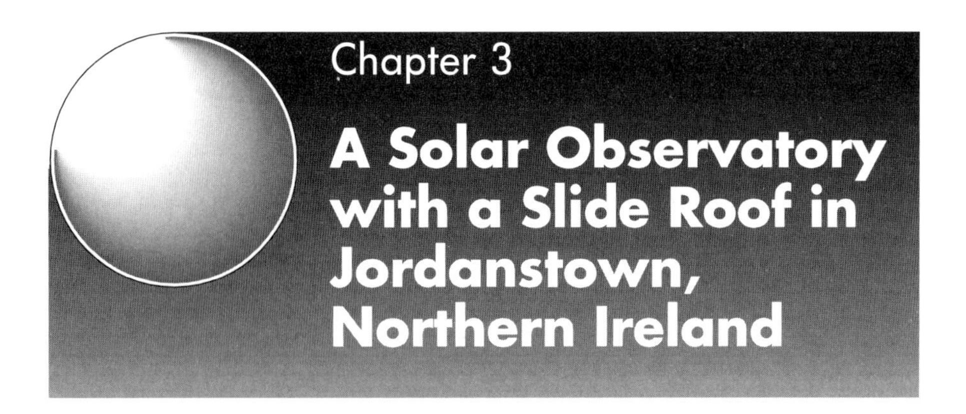

Chapter 3

A Solar Observatory with a Slide Roof in Jordanstown, Northern Ireland

Bruce Hardie

Figure 3.1 Bruce Hardie's solar observatory at Jordanstown.

Some years ago I needed to replace the run-off shed that covered my 130 mm (5.1 in) f/15 refractor. As I am almost exclusively a solar observer I wanted to meet two basic criteria. The first is to have a clear southern view of the sky from east to west. The second the ability to observe at the eyepiece end in a darkened observatory. Daytime seeing at my site in Jordanstown, Co. Antrim, was average to good, the open water of Belfast Lough lay 800 metres (half a mile) to the south-east and morning observations in that direction were generally good and sometimes excellent.

With these considerations in mind I decided against a dome with a slit as I thought that it would probably affect the seeing by trapping daytime summer heat within the structure. I therefore decided on an observatory with a slide-off roof.

I finally decided to adapt a garden shed of standard design.

The shed was model 810 Yardmaster made by Yardmaster International in Stockport, England. It is made of hot-dipped galvanised steel sheet with a paint finish baked on, so that maintenance is negligible. It should be noted that this particular shed does not come ready made up; like most garden sheds it is in a number of panels that have to be put together. You can collect it from the retailers yourself, and it will fit into a family hatchback or onto a fairly substantial roof rack.

Although full instructions come with it to make a garden shed, some fundamental changes have to be made to turn it into an observatory (see Figure 3.2). None is particularly difficult but each needs a little thought. In the "garden shed configuration" the roof is an integral part of the shed walls and the doors are hung from it: this of course has to be changed.

The basic method of construction that I used was to mount the shed on a course of breezeblocks (concrete blocks) to increase its height. A wooden frame was mounted on the bottom of the roof (now separated from the shed walls), and another wooden frame mounted on the top of the walls holding them together. The frames are made of 100 mm × 50 mm (4 in × 2 in) timber. I mounted the top runner for the sliding doors on the top frame which secures the walls of the observatory; previously the runner and the doors had been part of the roof assembly when in the "garden shed configuration".

I made up the roof into a separate unit attached to its own wooden frame.

I mounted castors on the roof frame to run on the wooden frame on top of the observatory walls. To the observatory I added a 75 mm × 75 mm (3 in × 3 in) timber gantry to support the roof when it was slid back. Along the gantry and continuing along the wooden frame on top of the observatory walls I fitted some aluminium channels, in which the roof castors ran. I added a brass curtain rail along the rear end of

Figure 3.2 (*opposite*) The plans for changing the garden shed into an observatory.

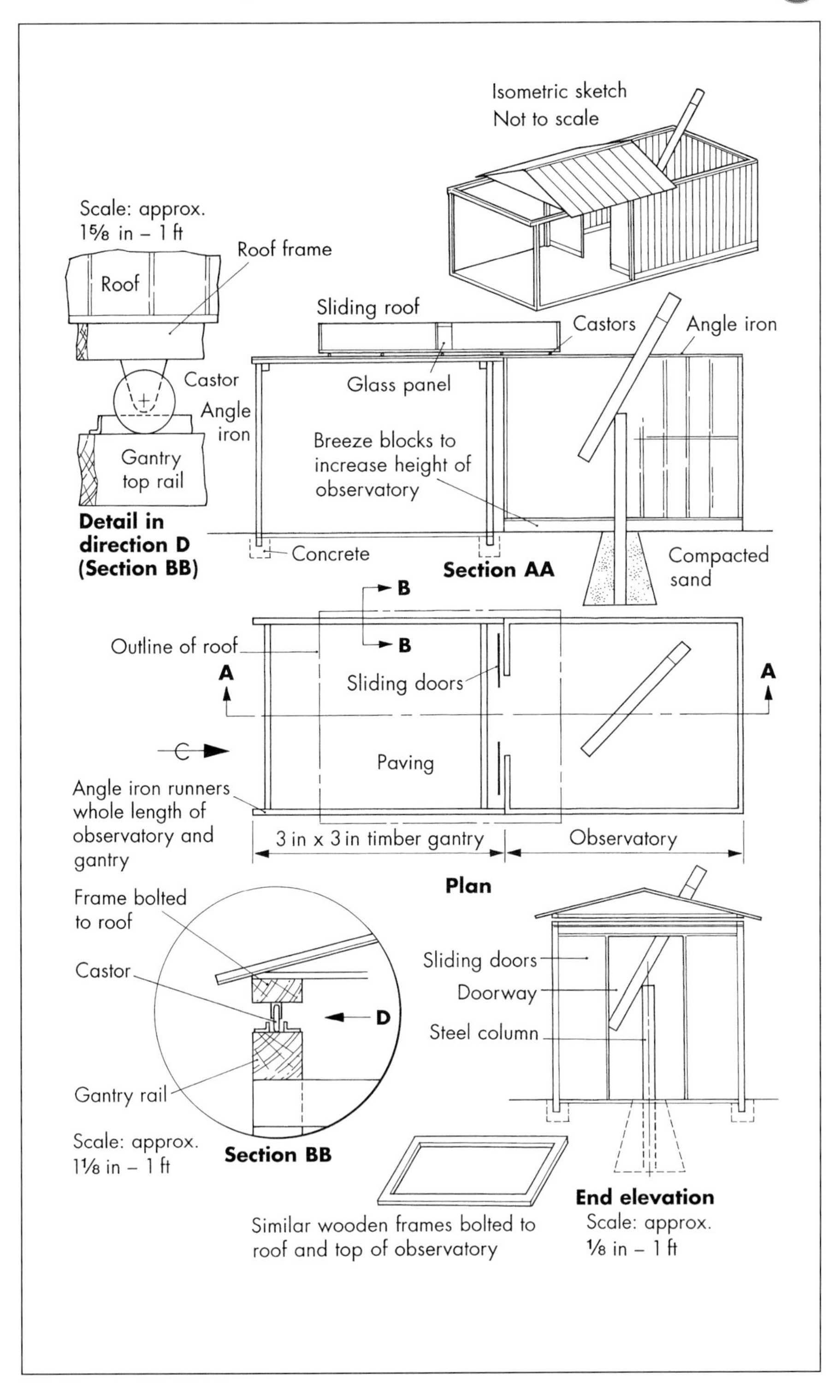
Isometric sketch
Not to scale
Scale: approx.
1⅝ in – 1 ft
Roof frame
Roof
Sliding roof
Castors
Angle iron
Castor
Glass panel
Angle
iron
Breeze blocks to
increase height of
observatory
Gantry
top rail
Detail in
direction D
(Section BB)
Concrete
Section AA
Compacted
sand
B
B
Outline of roof
A
A
Sliding doors
C
Paving
Angle iron runners
whole length of
observatory and
gantry
3 in x 3 in timber gantry
Observatory
Plan
Frame bolted
to roof
Sliding doors
Castor
Doorway
D
Steel column
Gantry rail
Scale: approx.
1⅛ in – 1 ft
Section BB
End elevation
Scale: approx.
⅛ in – 1 ft
Similar wooden frames bolted to
roof and top of observatory

the roof frame, opposite to the doors, on which a dark double curtain is hung. This ensures that the eyepiece end and the projection board remain in a darkened area when observations of the solar disk are being made.

The overall size of the observatory is approximately 3 m × 2.4 m (10 ft × 8 ft).

Before I assembled the observatory I had selected a site sheltered by a north-facing hedge which kept the prevailing winds off. I then marked and dug out the foundations and cemented in the breezeblocks. While these were settling in I marked and dug a hole for the telescope mount; I mounted the telescope slightly off-centre to the south.

The mount was a steel pipe buried in sand 1 m (3 ft) deep. I dug a wedge-shaped hole, wider at the *bottom.* After I put the pipe in I used compacted sand to secure it. I poured dry sand in layers about 150 mm (6 in) deep, tipping water on each layer, until it reached the top of the hole. The steel pipe was also filled with sand. Once this was done I assembled the observatory walls, attaching them by the bottom metal frame to the breezeblocks; this is a two-man task.

The floor is wooden which allows some measure of safety when dropping your favourite eyepiece! I cut a hole in the part of the floor adjacent to the telescope mounting to keep it vibration-free.

I have used the observatory for about fifteen years now under all sorts of weather conditions. The only time that I cannot use it is during gale-force winds – in such conditions the open roof is liable to take off.

When the observatory is not in use large carriage bolts between the wall and roof frame are passed through large eyelets and hold the roof secure. Although there is a gap between the roof and the walls, I have not covered it in as I find it helps ventilation, and no rain gets in because the roof eaves overlap.

Using the Observatory

As I said at the beginning of this chapter, I use the observatory primarily for solar observation, by the projection method for white-light work, and by direct viewing for H-alpha observation with a DayStar filter. I also undertake photography in both disciplines.

I have found a few problems.

I would perhaps like a little more space to work in. During winter mornings the sun is very low at my northern latitude and the top of the telescope tube hits the top of the walls: I have to wait until the Sun gets higher in the sky. I thought of mounting the telescope higher within the observatory, but then it would foul the slide-off roof. I suppose I could make some alterations to the roof height by building up the sides and ends of the roof structure itself.

Seeing has not been affected by the observatory structure. The observatory is surrounded by grass – no concrete paving (which would retain and then radiate heat) is used, and the path leading to it is made from small greyish quarry stones.

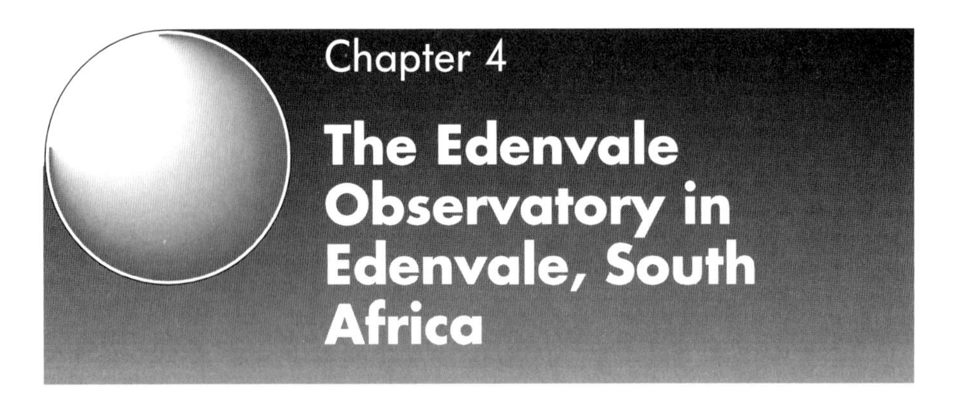

Chapter 4

The Edenvale Observatory in Edenvale, South Africa

M.D. Overbeek

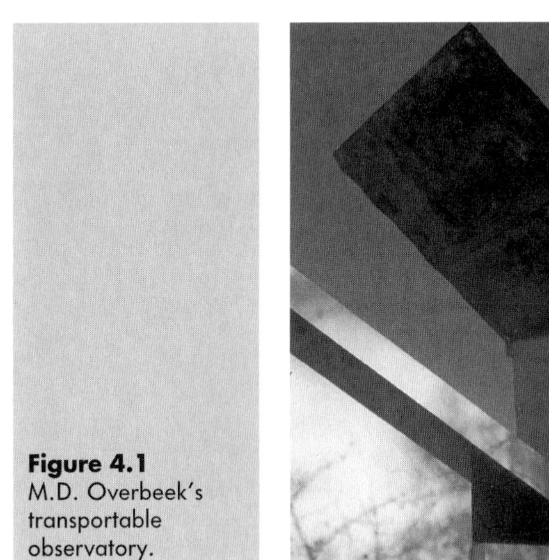

Figure 4.1
M.D. Overbeek's transportable observatory.

The Telescope

This chapter is meant to encourage those amateur astronomers who do not have the space, means or time to build an observatory like some of the splendid examples described in this book. If you are an amateur and are burning to get down to some serious work, you probably feel a need of something better than a site on the back lawn where street lights have to be dodged and other hazards of portable telescope observing coped with. Remember, all you need from an observatory is protection for the telescope from the weather and a handy observing base where you can become operational at very short notice.

The rest is vanity, as Hamlet should have said.

In the early 1950s, after making a thousand or two variable star and occultation observations from a sometimes windy and dusty back lawn, I decided that serious amateur observing deserves better conditions. At that time I was using a completely home-made 6 in (150 mm) Newtonian – even the eyepiece elements had been ground and mounted by an amateur. The telescope was convenient to use but a number of faint variables were beyond its light grasp. Here was a good case for a larger instrument, protected from the elements. The telescope had to have a reasonable light grasp, be highly manoeuvrable and had to be affordable. Large off-the-shelf catadioptrics lay some years in the future and so my instrument had to be home made.

I duly set out to build a $12\frac{1}{2}$ in (317 mm) Dall-Kirkham Cassegrain telescope, and about a year later the instrument was in use. The optics proved to be straightforward to make. The final Foucalt tests of the ellipsoidal primary employed a small pinhole source at the near conjugate focus and a knife edge at the far conjugate focus. The geometry of this arrangement made it a true null test. Then I partly polished and figured the concave tool of the secondary to a spherical shape and used it as a test plate for Newton's fringe testing of the secondary.

The optics were mounted in a very rugged reinforced square Masonite hardboard tube, just 48 in (1.2 m) long. I gave the optical tube an adequate equatorial mounting with a clock drive, large, easy-to-read setting circles and smooth slow motion controls. The instrument has a small finder and a 3 in (75 mm) auxiliary refractor. A parfocal turret gives a choice of 167 or 267 diameters through the main telescope, and the refractors have magnifications of 6 and 48 diameters. As the various eyepieces and controls are all within a radius of about 10 in (250 mm) I can switch between these four powers within a second or two. It takes but little longer to change from one familiar variable star field to the next.

Two Observatories

Before going into details, I would like to air a few thoughts on amateur observatories.

If you plot the amateur observing stations around the world on a graph with total cost in material and labour along the horizontal axis, and annual output of scientific data along the vertical, you will find a strong negative correlation between the two.

At the one extreme is the world's most prolific and accomplished variable star observer, Albert Jones OBE of Nelson, New Zealand. Albert's $12\frac{1}{2}$ in (317 mm) home-made Newtonian is kept in a tool-shed from which he wheels it for a night's observing. His instrument is strongly reminiscent of the principal industry of his country, which is agriculture.

At the other extreme are any number of beautifully built and equipped observatories, whose builders are too busy perfecting their facilities to get down to serious observing. Aspiring serious observers will do well to ask themselves what they really want from their observatories.

My own interest is in so-called serious amateur work, that is the making of observations that are reported formally and used by the professional astronomical community, in this case variable-star observing and occultation timing. I have no quarrel with recreational astronomers who observe the heavens for the sheer pleasure of it or those who take astronomical photographs which are afterwards admired but not used for scientific purposes. These good folks pursue different goals and I am not qualified to speak for their needs. Amateurs who are dedicated to the bringing of astronomy to the public are in a class of their own and I cannot really speak for their needs either. It is probably better not to confront newcomers with beautiful, expensive facilities but on the other hand a beginner who is asked to look through a telescope which keeps on losing the object, on a windy and light-polluted back lawn is also likely to be discouraged.

The First Observatory

The building in which my own $12\frac{1}{2}$ in (317 mm) reflector was first housed was adapted for that purpose at no cost apart from three 50 mm × 100 mm × 190 mm (2 in × 4 in × 6 in) pine studs and a handful of hardware. I had designed the telescope and mounting to fit into the end of my garage which has a gently sloping corrugated iron roof (see Figure 4.2).

It was a simple matter to remove about 2 m^2 (20 ft^2) of sheeting and to make a panel by fastening the removed sheets to two of the studs. The studs straddled the hole left by the removal of the sheets and I attached wooden runners made from the third stud to their four ends. These runners lay in the troughs of the undisturbed sheets and allowed the panel to slide easily. Two hardware store pulley sheaves enabled me to open and close the panel by pulling on a rope. The alterations weakened the roof structure slightly and probably violated any number of clauses in the municipal building code but then I never consulted the powers that be.

Figure 4.2 The first observatory, showing the sliding panel.

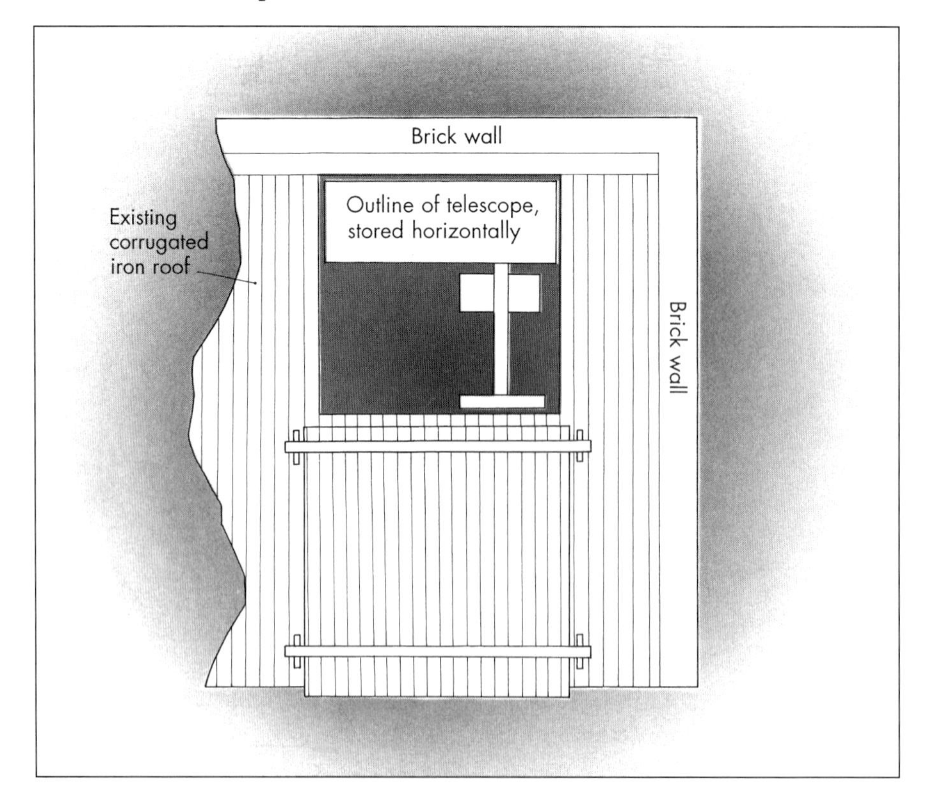

If you do want to consult someone then it is a good idea to let your neighbours know what you are planning. If they have enjoyed a look through your telescope, they will probably be supportive when it comes to yard lights on their side of the fence, and electronic noises from your side in the small hours. In this connection I must digress for a moment to tell of an interview with a house owner while we were house hunting. I asked him about street lights and he replied, "Don't worry, it is as bright as day here all night."

To return to my first observatory: the whole job took a Saturday and was completely successful. The horizon was about 25° all round. The telescope was used for many occultation timing and variable star observations as well as for obtaining thousands of photographs of Mars during the favourable 1956 opposition.

The Present Observatory

As we learn to our sorrow, the good things of life do not always last. Domestic circumstances dictated a move to another locality and I had to start again, without the benefit of a garage with a corrugated iron roof. Having had to move repeatedly, I was not keen to build a permanent observatory and opted for a temporary, transportable one. The decision proved to be sound. The building described below has in fact been moved three times since it was completed.

It often happens that temporary measures have to serve permanently and this observatory, after 23 years, is a case in point. The photograph (Figure 4.1) shows that it is no longer the trim structure that was built in 1972. Rough handling during the moves and failure on my part to do maintenance have taken their toll. The footing has been badly damaged by rainwater because I omitted to erect it on a course of bricks after its last move. Nevertheless, the building is still weatherproof and it continues to serve its purpose.

The building consists essentially of four panels of Masonite hardboard, 3 mm ($\frac{1}{8}$ in) thick and stiffened with bracing pieces at 600 mm (2 ft) intervals. The panels are bolted together and are capped by a hinged roof of the same material. The roof has a radius,

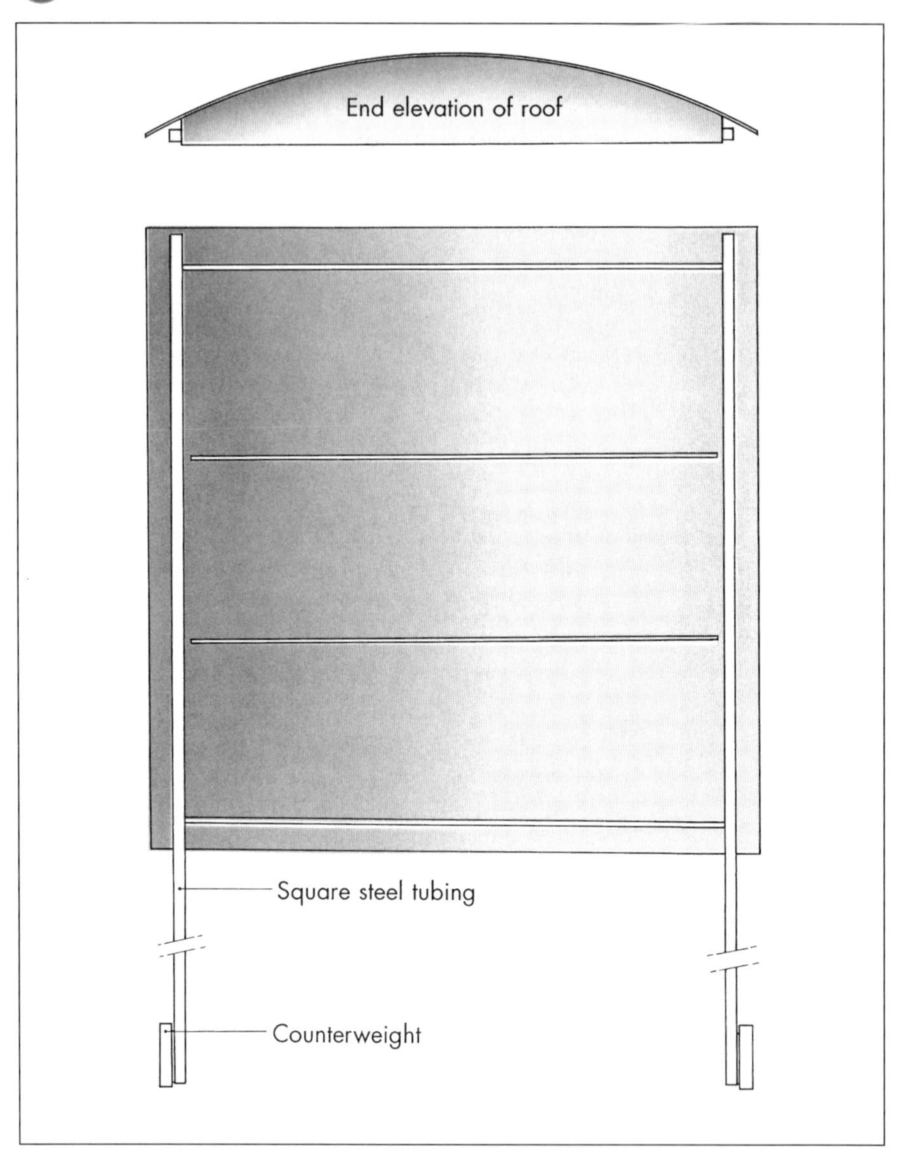

which imparts a surprising amount of stiffness to it (see Figure 4.3). So far it has withstood the onslaught of the large hailstones which are a feature of our South African Highveld summer. The roof is counterweighted and is easily operated with one hand. The telescope is mounted on a 200 mm (8 in) steel pipe with a 600 mm (2 ft) diameter flange which rests on the flat concrete roof of an outbuilding (see Figure 4.1). Vibration is not a problem.

Figure 4.3 Underside of the roof showing endpieces and stiffeners.

The observatory door is only five horizontal, and ten vertical steps from my kitchen door. It takes less than a minute to leave the kitchen, open the observatory door and roof, uncover the telescope and switch on the drive and chart light. Reversing the procedure also takes less than a minute. The importance of having an observing station which is convenient in every respect cannot be over-stressed. Amateurs are not paid to spend nights observing, but they can produce valuable results by spending short sessions at the eyepiece, sometimes between clouds and sometimes between domestic activities. A super-convenient observing facility is a great incentive to go out and do useful work. In my opinion a modest but convenient-to-use telescope is far superior to an expensive but inconvenient facility for producing significant results. The large, old fashioned setting circles on my instrument are a case in point. My telescope can be pointed at the desired field more rapidly than expensive computer assisted instruments.

Other Equipment

I use a home made 1P21 photoelectric head for observing occultations of bright stars by the Moon or minor planets. The recording is done on a high speed strip recorder. It takes only minutes to convert the telescope to the PEP mode.

A Julian Day clock based on a 1951 *Sky And Telescope* article by Frank Bradshaw Wood gives the last two integers and first four decimals of the day. By using this clock at the telescope, I eliminate tedious conversions when making up reports.

A two-channel home-made seismograph records powerful earthquakes from all over the world, as well as earth tremors that are triggered by mining operations up to several hundred kilometres away. A solar flare detector which utilises the sudden ionospheric disturbance phenomenon, and a magnetograph for detecting flare-related magnetic disturbances complete the list of scientific equipment. The recording from these three devices is not done in the observatory building.

Second Thoughts

The aluminium coating of the telescope's primary mirror deteriorates rapidly in the polluted atmosphere where I work. The fact that the instrument is pointed at the sky for many hundreds of hours a year does not help. If I had realised just how extensively the telescope would be used, I would have taken a deep breath and invested in a 14 in (350 mm) Schmidt-Cassegrain optical tube when these became affordable.

Also, if I had known that the observatory would still be in use twenty-three years after it was built, I would have done more to protect it from the elements. I would also have made it just a few inches longer, because it is difficult to observe a few northern hemisphere stars, which were added to my repertoire after the observatory was built.

Small Observatories

Readers should be aware by now that I have a penchant for small, easily affordable observatories. My observatory, although small, dwarfs the one used by variable star observer Eric Harries Harris of Adelaide, South Australia. Eric's structure just manages to enclose his 6 in (150 mm) Newtonian. It is so compact that he could not find room to mount the observatory clock which is now perched on the end of the telescope like a kookaburra bird!

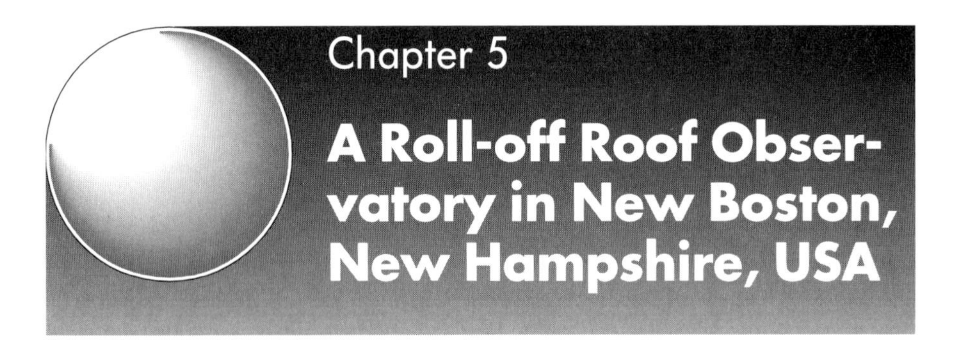

Chapter 5
A Roll-off Roof Observatory in New Boston, New Hampshire, USA

Lawrence D. and Linda Lopez

Figure 5.1 Lawrence and Linda Lopez' roll-off roof observatory.

This is the third observatory we have built since we purchased our first telescope in 1983. It is 12 ft × 16 ft (3.6 m × 5 m) with a roll-off roof. We decided to build the observatory ourselves due to money considerations – we had a new mortgage and had just ordered an Astrophysics 7 in (178 mm) f/9 refractor.

We started in August 1992 about a year after the house was finished. Total construction time was about forty days, and the telescope was installed in May 1993.

We currently live on 18 acres in New Boston, New Hampshire, USA (latitude 43°, longitude 71°). We chose the land because it was private, fairly dark with low horizons, and was within commuting distance from possible jobs.

The hilltop site presented difficulties that had to be addressed in the design. The ground drops about 8 ft (2.5 m) from north to south over the 32 ft (9.75 m) of the structure. Pier supports seemed to be the easiest way to deal with the slope and they allow air circulation around and under the building to speed cooling. The observatory is exposed to strong winds and subject to heavy snow accumulation, so the roof has to be strong and weather-tight, but light enough to move around with a foot or two of snow on it.

A Newtonian reflector was not considered for the observatory because of wind exposure. Also, since the design was centred around a refractor the walls could be higher without interfering with the telescope. The 7 ft (2 m) walls shield us from wind (important in January), lights and some of the biting insects of summer. There also is room to stand upright with the roof closed.

We located the observatory about 100 ft (30 m) from the house. This is close enough to be accessible, even through a couple of feet of snow. The proximity also allowed us to run extension cords to the site for tools and later for equipment in the observatory.

The Building

The building (see Figure 5.2) is supported on the east and west sides by three 10 in (250 mm) diameter concrete piers. Two additional piers to the north support the ends of the run-off rails. The length of the piers varies from about 5 ft to 9 ft (1.5 m to 2.75 m) due to the slope. A 2 ft 6 in (760 mm) diameter concrete base pier for the telescope mount is located in the centre of the building. The footings for the piers are set 4 ft (1.2 m) below ground to go below the frost line. Excavation work took two days using a small backhoe. Iron reinforcing bar was used to tie the footings to the piers, and J-bolts set into the concrete were used to attach the building to the piers. The support piers used a total of 2900 lb (1318 kg) of concrete. The base for the telescope pier was poured by truck, using 1.5 cubic yards (1.15 m^3) for the 7.5 ft (2.3 m) length.

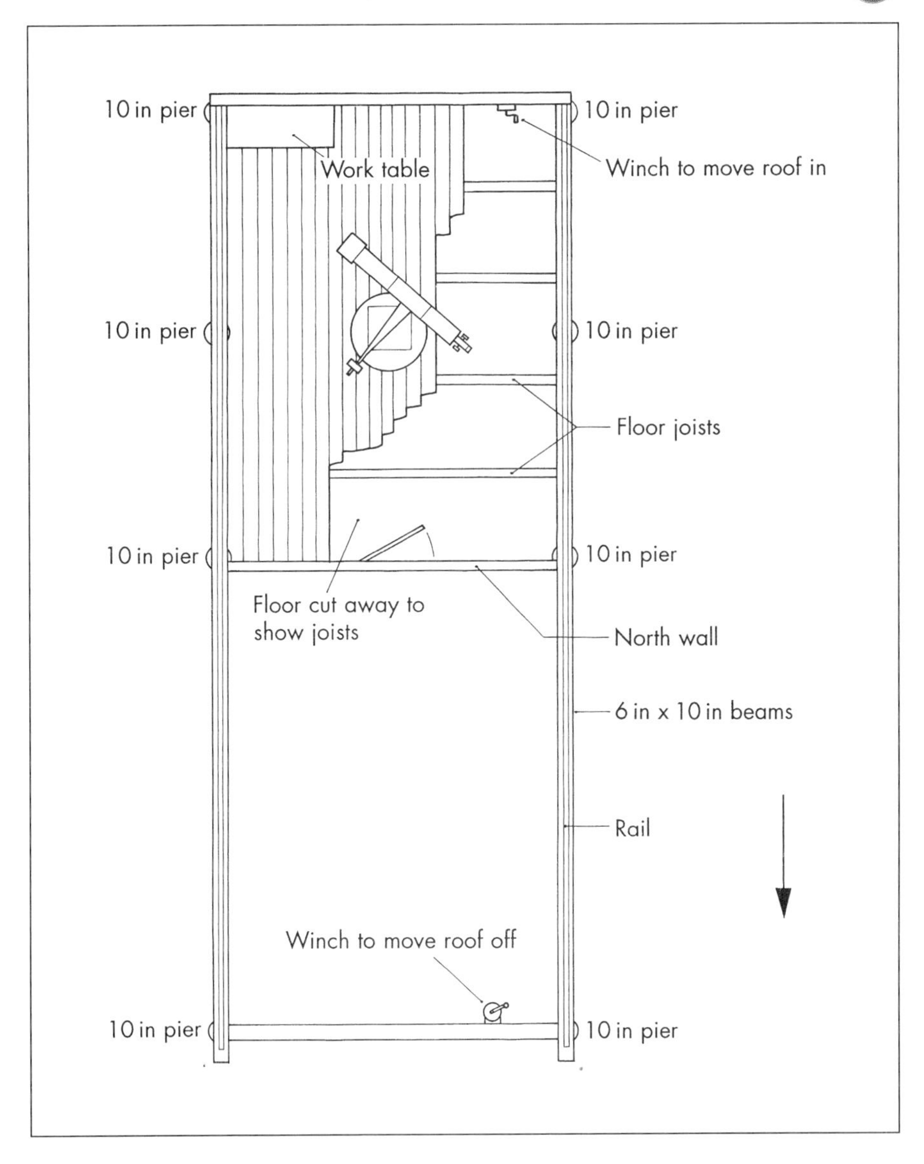

Figure 5.2 Top view of the observatory without the roof.

The floor support consists of two 6 in × 10 in × 16 ft (150 mm × 250 mm × 4.9 m) beams and six 4 in × 10 in × 12 ft (100 mm × 250 mm × 3.65 m) floor joists. Pockets cut in the top of the beams hold the joists. Pine boards were used for the flooring. The large rough-cut timbers were inexpensive local products.

In the walls, 2 in × 6 in (50 mm × 150 mm) construction provides strength and an adequate base for the run-off rails. The walls are covered with exterior plywood and vertical shiplap siding.

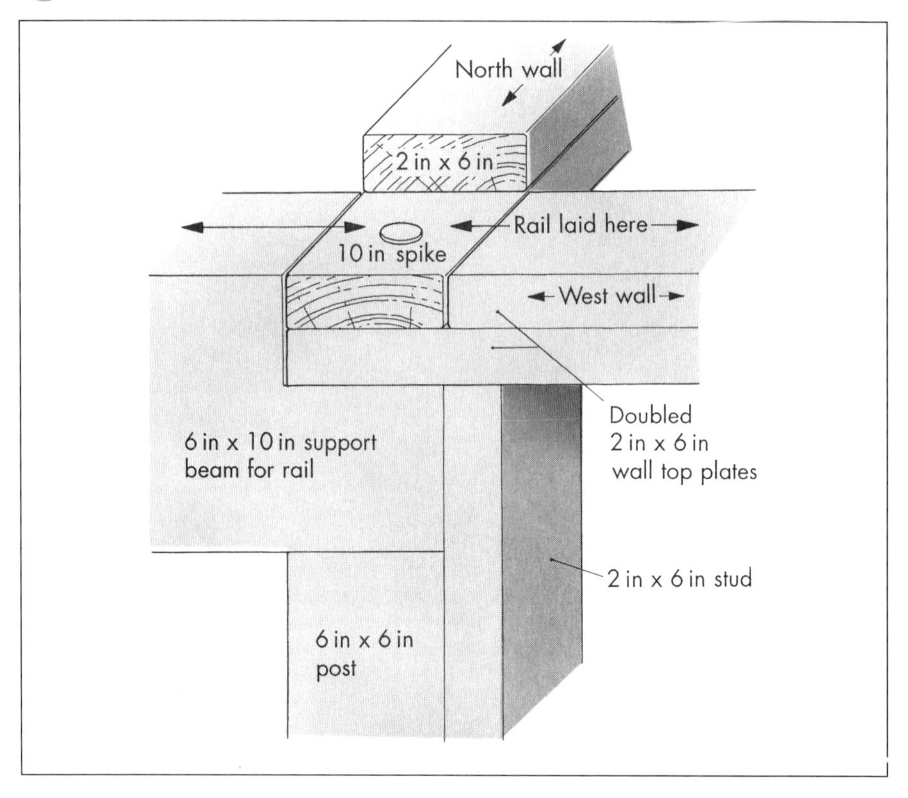

Figure 5.3 Detail of the northwest corner.

The wall is standard 2 in × 6 in (50 mm × 50 mm) construction. The top plates of one wall form lap joints with the top plates of the next wall. To accommodate the run-off rails, the studs of the north wall sections were designed to leave a 6 in (150 mm) square gap in the corners. 6 in × 6 in (150 mm × 150 mm) posts were placed in the north corners and nailed to the floor and the wall studs on either side. These posts were cut short to leave a 6 in (150 mm) gap to the lapped top plates. The notched ends of the support beams for the run-off rails were inserted into the gap to rest on the 6 in × 6 in posts and secured by 10 in (250 mm) spikes put through the top plates and beams and into the end of the post (see Figure 5.3).

The Run-off Roof

Two inch (50 mm) angle-iron welded face down on a 5 in (125 mm) flat plate forms the rails, a design that does not accumulate snow and ice or channel water

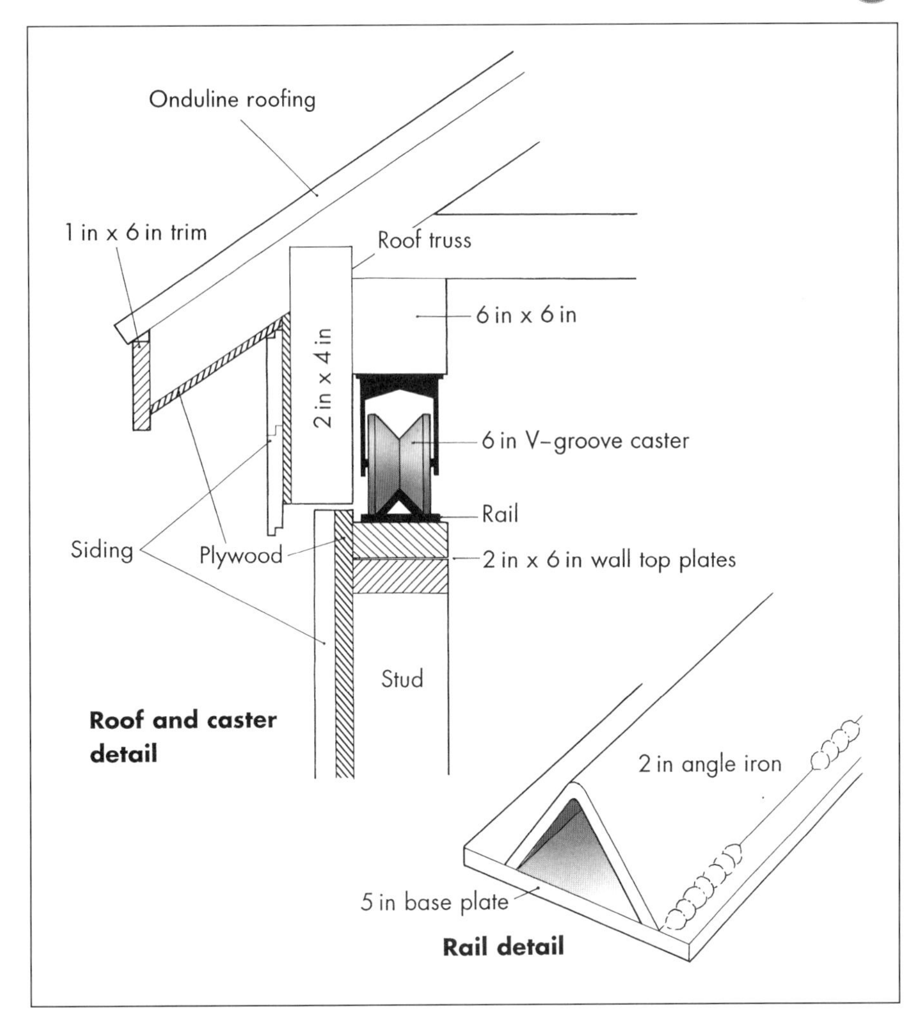

Figure 5.4 Details of the roof, casters and rail.

into the building (see Figure 5.4). Each side uses three 10 ft (3 m) sections. The centre section of rail overlaps the join of the walls and the support beams to further tie the structure together. The far ends of the support beams rest on 6 in × 6 in posts set on two concrete support piers. The posts are tied together with a 6 in × 6 in beam and stabilised by knee braces.

The winch to move the roof off is mounted on the cross-beam. Another winch is mounted inside, on the south wall to pull it back on again. An extension of the south wall and blocks nailed to the north ends of the rail support beams keep the roof from rolling off.

The roof frame uses nine 2 in × 4 in (50 mm × 100 mm) trusses mounted on a carriage made with two 16 ft long pieces of 6 in × 6 in (150 mm × 150 mm), tied together with 2 in × 4 in (50 mm × 100 mm) lengths.

The carriage was constructed in position, since it would have been too awkward and heavy to lift as a unit. Three 6 in (150 mm) diameter V-groove casters are positioned on the rail under each 6 in × 6 in (150 mm × 150 mm) so that they are over the building support piers when the roof is closed. Corrugated asphalt roof panels are nailed to stringers across the trusses. The panels are easier to install than conventional roofing and reduce the weight.

Since the casters raise the roof $8\frac{1}{2}$ in (216 mm) off the walls the gap had to be closed to keep out the weather. We therefore installed plywood and siding under the edge of the roof, on vertical sections of 2 in × 4 in (50 mm × 100 mm) attached to the outside of the carriage. The siding on the north and south ends leaves only a quarter-inch (6 mm) gap between roof and walls and so is not a problem.

We then used the finished roof as a crane! It was used to move the steel mounting pier from the truck to inside the observatory. The pier is $\frac{3}{8}$ in steel pipe, 1 ft 4 in (400 mm) in diameter and weighs 500 lb (227 kg) (see Figure 5.5). It is attached to the concrete base by a 4 in (100 mm) welded flange and eight $\frac{3}{4}$ in (18 mm) J-bolts set into the concrete. A 1 in (25 mm)

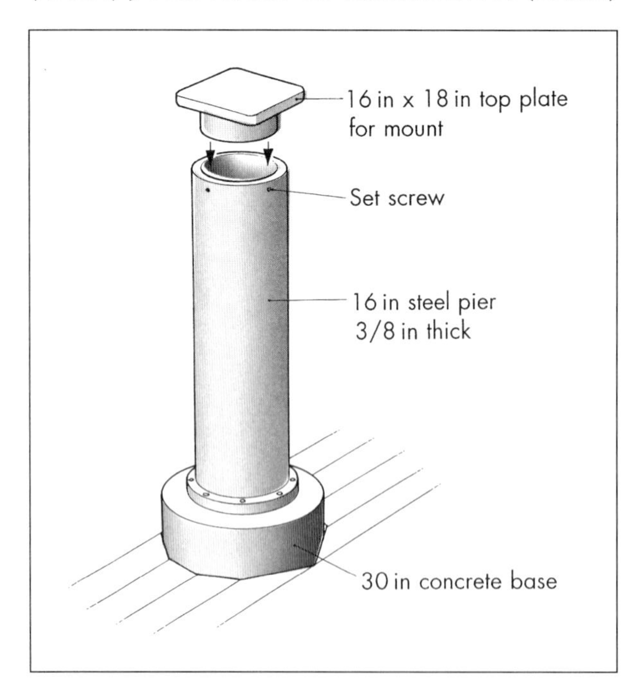

Figure 5.5 Pier details.

steel plate, 1 ft 4 in × 1 ft 6 in (400 mm × 456 mm) welded to a collar sits inside the top of the pipe. The corners of the plate were rounded to preserve our scalps.

A Byers 812 mount is bolted to the plate. Polar alignment is carried out by rotating the plate, securing it with set screws, and then setting the altitude on the mount. Since the pier assembly is isolated from the rest of the structure there is little vibration.

Looking Back

This observatory fulfils the original design requirements well.

The equipment is always ready to use and requires little adjustment. We are shielded from wind, light, and bugs, and there is enough room to move around the telescope and to accommodate three or four guests as well as a worktable. We need a ladder only for objects close to the horizon.

The roof moves easily, even under eighteen inches of snow – maybe too easily! We came home one day to find a strong wind had blown the roof 4 ft (1.2 m) out onto the rails because we did not keep the winch engaged.

And the building was relatively inexpensive at about $2500, modest for a building of this size.

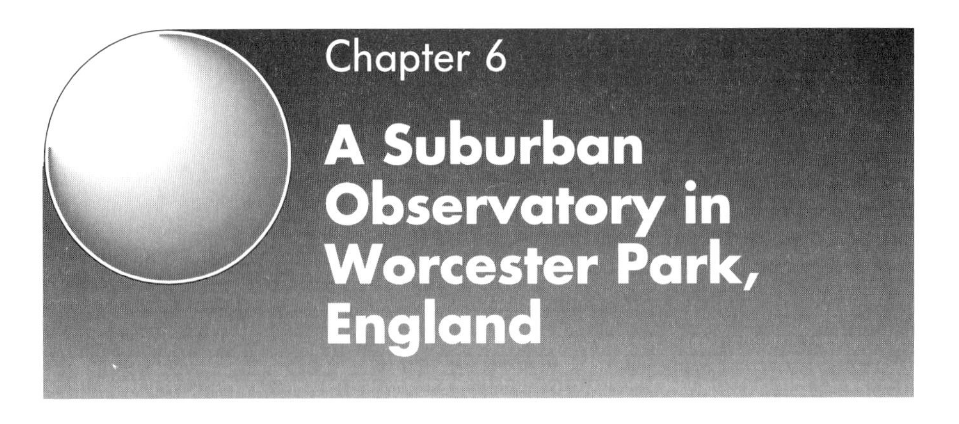

Chapter 6

A Suburban Observatory in Worcester Park, England

Maurice Gavin

Figure 6.1 Maurice Gavin and his twin observatories.

Because of the vagaries of the British climate I have found that more observing can be done from my own back garden than by moving equipment to a remote and maybe darker site. In any case, transportation becomes pretty impractical for an equatorially driven telescope of over 250 mm (10 in) aperture – and I had a 440 mm ($17\frac{1}{2}$ in) Newtonian in mind.

Every observing session begins by checking the clarity of the sky. It can be deceptive. Most solar system objects are bright, and steady seeing is more important than crystal clarity. A town dweller is not disadvantaged in this respect. Even if only a sprinkling of stars down to 3rd magnitude are visible (bad enough to send more favoured observers indoors), the telescope will still reveal very faint stars. This is especially so if a modern electronic camera is used. I have clearly recorded – in a few seconds – onto a CCD camera both the elusive Horsehead nebula and the central star in the Ring nebula in late twilight. Even from the darkest site with a large Dobsonian these objects can remain completely invisible to the eye.

Such an electronic camera and associated equipment should not be regarded as expensive, for savings can be made by using a more modest telescope in a compact observatory. CCDs effectively increase a telescope's penetrating power maybe a hundredfold, and turn a small instrument into a telescope only a professional could afford if a CCD camera were not used. Imagine: an amateur's 250 mm aperture telescope matching the light-collecting power of yesteryear's Hooker 100 inch (2.5 m) telescope!

A Rotating Dome

There are various options for a home based observatory. I chose to build an observatory with a rotating dome (see Figure 6.1).

It would provide complete protection for myself and for the telescope, and would be permanently set up and immediately ready for use. Wind buffeting of the telescope is minimised in a dome – important for my interest in long, guided photographic and CCD exposures. Dewing-up of the telescope or cameras can be eliminated by a domed building, particularly if building materials are carefully selected with this in mind.

During my observatory's twenty years of continual use, no dewing of the optics has ever occurred.

The telescope, mounting, and observatory were designed as a single entity. The declination axis was set high on the telescope tube to minimise the arc described by the Newtonian eyepiece, while the whole mount was to be set as low as possible into a cut out in the floor.

I made it a high priority to be able to reach the eyepiece in any position without using steps and this was successfully achieved. The massive forks could accommodate my proposed 440 mm ($17\frac{1}{2}$ in) aperture Newtonian with smaller instruments clustered about to counterbalance it without stress.

I settled on a 3 m (10 ft) diameter dome with up-and-over shutter as being the minimum practical size with sufficient clearance for both telescope and observer. I did consider alternatives, and arrived at what I thought was the best size. For example, increasing the dome radius by just 300 mm (12 in) increases the surface area (and potential cost) by 44%. Conversely a dome only 1.7 m (5 ft 6 in) diameter (see p. 52) has only a third of the surface area of a 3 m (10 ft) diameter dome.

With my location in the London suburbs, light pollution from general skyglow and neighbours' security lighting can be severe. Often the "up-and-over" shutter is opened just sufficient to frame the celestial object under scrutiny. Even so, its width is a full third of the dome diameter, 1 m (3 ft 3 in) wide. This lets me view the sky from beside the shutter opening without blocking light to the telescope, and also reduces the need to continually rotate the dome to track a target. A wide shutter also allows a wide-angle camera perched on the end of the telescope tube to take panoramic views of the sky without vignetting from the dome. I produced a simple photographic all-sky star-atlas this way.

The Observatory Building

Plans were drawn up to for an economic structure that could be made with hand tools plus a power drill and jigsaw. I used full sheets of standard 2.4 m ×

1.2 m (8 ft × 4 ft) composite board or plywood for the flat roof and dome structure, with a requirement for the minimum number of cuts and waste dictating many of the dimensions.

The observatory takes the form of a rectangular domestic garage. Located at the bottom of my garden, remote from the house, it has generally unobstructed views of the entire sky. The building measures 5.2 m × 2.5 m (17 ft × 8 ft) internally, and is 2.4 m (8 ft) high with full-width folding doors at one end and a large window and access door facing the garden.

It initially served as a workshop to build the dome and telescopes. The flat roof is oversized to accommodate the dome at one end, over a raised 1.3 m (4 ft 4 in) high observing platform with storage space below. The observing platform is covered with an old carpet which protects the occasionally dropped eyepiece.

The remaining space at ground level forms a 2.5 m × 2.2 m (8 ft × 7 ft) office/workshop. Two sides are lined with worktops built from discarded flush doors on 900 mm (36 in) high drawer units. A small wardrobe with a 60W lamp in a metal box serves as a heater, keeping maps and eyepieces dry. Opaque roller blinds of black vinyl sheet can be dropped across the window and observing platform to isolate the office area so as to make a temporary photographic darkroom.

Building Construction

The London clay subsoil cuts like firm butter, but is prone to shrinkage or frost-heave to a depth of 1.2 m (4 ft) in severe seasons. This can disturb foundations, including the alignment of the telescope pier.

I hand-augered holes 200 mm (8 in) in diameter and 2 m (6 ft 6 in) deep around the perimeter and filled them with concrete and steel reinforcement in a form of "piling". These were then linked at ground level with a 200 mm × 100 mm (8 in × 4 in) wide reinforced concrete edge-beam, cast *in situ*. This greatly reduced the material, labour, and cost involved where virtually every spade-full of soil excavated must be replaced with concrete. Clearly the foundations must suit local conditions, and mine, of course, are not universal.

The walls were then raised off the edge-beam in standard 100 mm (4 in) thick concrete blockwork.

The roof is of standard timber construction using second-hand joists, trimmed around the dome aperture and clad in 20 mm ($\frac{3}{4}$ in) composite building board and two layers of bituminous roofing felt with a mineral finish. The roof has a crossfall of 75 mm (3 in) for drainage and the dome also rotates on this inclined plane without problems. Piling was used for the separate telescope pier foundations, raised to the observing platform level in concrete blockwork and capped in concrete with bolts cast in to receive the equatorial head.

A masonry building is said to act as a heat store, giving off radiation at night and spoiling local seeing. I haven't observed these effects myself. The potentially troublesome long south observatory wall is shaded by a high timber fence while the east-west elevations are either of timber or shaded by planting.

Dome Construction

A variety of dome shapes can be built of flat or curved sheet material. What seems vital, in my experience, is that the base of the dome is completely rigid and that the dome walls are near perpendicular at this abutment.

Initially I proposed and planned a 3 m (10 ft) fibreglass dome – see Ron Johnson's observatory in Chapter 11. Unfortunately materials prices rose dramatically just at the wrong time, making that option uneconomic, so I built a plywood dome instead.

My plywood dome has no internal ribs and gets its strength from a rigid box beam or base-ring at the perimeter and two vertical plywood hoops supporting the up-and-over shutter (see Figure 6.2). The continuous curved box beam is made of marine plywood – the top and bottom 12 mm ($\frac{1}{2}$ in) members are separated with 2.5 mm ($\frac{1}{8}$ in) ply sides and lined internally at 200 mm (8 in) intervals with 12 mm ($\frac{1}{2}$ in) thick softwood blocks called diaphragms. When glued and nailed together (with all the joints staggered) it forms an extremely strong but lightweight structure.

I used a similar double-walled tube construction for my main Newtonian telescope.

Figure 6.2 The 1.7 m Meade dome; the 16-sided plywood frame is clad in sheet aluminium.

Using a pocket calculator to compute its shape, I cut a master gore as a template. The gore should not be too wide at the base – any material in sheet form resists being bent in two planes at once, something that occurs in the dome where it meets the base-ring. By tacking several full sheets of 2.5 mm ($\frac{1}{8}$ in) marine ply together, several gores can be cut simultaneously. The dome required sixteen in all, two being cut to form the make-up pieces abutting the vertical shutter hoops. Each gore, temporarily curved over a plywood rib of correct radius, was stitched to its neighbour with copper wire at regular intervals and nailed to the base-ring and shutter hoops.

The dome is weather-proofed externally with discarded 0.3 mm ($\frac{1}{100}$ in) thick A3 size (297 mm × 420 mm) aluminium printing plates, epoxied to the ply dome skin and sealed along the gore butts with Flashband – a bituminous, foil-coated strip used for sealing roofs. I painted the dome with aluminium paint, which reflects and takes up the sky colour to blend well into the skyscape. The interior of the dome is painted a pleasing matt blue.

The low conductivity of the timber construction appears to prevent convection currents across the shutter aperture, so there seems to be no need for the more often recommended funereal black interior. I paid particular attention to eradicating scattered light *within* the telescopes: all are of closed-tube construction, and have matt black interiors.

The dome rotates on five equally spaced heavy-duty nylon castors, inverted and bolted to the observatory flat roof. These bear and support the underside of the base-ring. The dome is kept on track by four equally spaced horizontal restraint rollers bearing onto the base-ring walls. These rollers include metal plates to stop the dome lifting in storm conditions.

Whereas the dome is hemispherical (formed by the petal-like cylindrical curves), I chose an up-and-over shutter of cylindrical section, made from 1.5 mm ($\frac{1}{16}$ in) thick rectangular aluminium sheet. This was curved onto a plywood frame of slightly greater radius than the dome itself for clearance. Such a shutter is easy to construct and seal from the elements; it hugs the dome profile and so does not act as a wind scoop when open. Also, the shutter aperture is fully controllable.

The shutter rides on four roller-skate wheels with four side-restraint rollers and can be opened beyond the zenith. The fixed back-section, over which the shutter rides, is made of curved aluminium sheet pop-riveted together. The shutter is not completely rigid, but flexes slightly along its length so that the wheels are always in contact with the dome roof. At the bottom of the shutter and fixed to the base-ring is a smaller section that hinges forward for observation along the horizon. Extruded aluminium sections interlock and seal the ends of the shutter when closed. I use a pole to raise the shutter. Chromium-plated chains hold it in position against gravity. The top of the shutter has a section of clear corrugated plastic acting as a roof-light and providing welcome daylight into the observatory interior.

Recent Developments

The equatorial fork mount, driven by an Irving 40cm (16 in) diameter worm and wheel, has a massive plywood cradle which has supported various home-made telescopes down the years (see Figure 6.3). Currently these include a 440 mm (17$\frac{1}{2}$ in) f/4.5 Newtonian, a 260 mm (10$\frac{1}{4}$ in) f/4 reflector as a dedicated CCD camera (no option for visual use), a Wray 230 mm (9 in) f/4 lens, and a 200 mm (8 in) Celestron C-8 SCT.

Figure 6.3 The main observatory interior under the 3 m dome contains a clan of telescopes around the 440 mm Newtonian (white tube).

A full-aperture objective prism (on British Astronomical Association loan) before the CCD reflector converts it into a stellar spectrograph. Increasing light pollution has rendered long, guided photographic exposures impossible, and CCD cameras have brought some respite. However, fast f/4 optics are prone to light pollution and the small size of CCDs makes finding faint objects a chore in the absence of setting circles or declination control.

A Second Observatory

With this in mind I purchased a fully computerised Meade 300 mm (12 in) f/10 SCT LX200 telescope in 1995. To exploit its potential, I built a second observatory to the west, abutting the original observatory (see Figures 6.1 and 6.4).

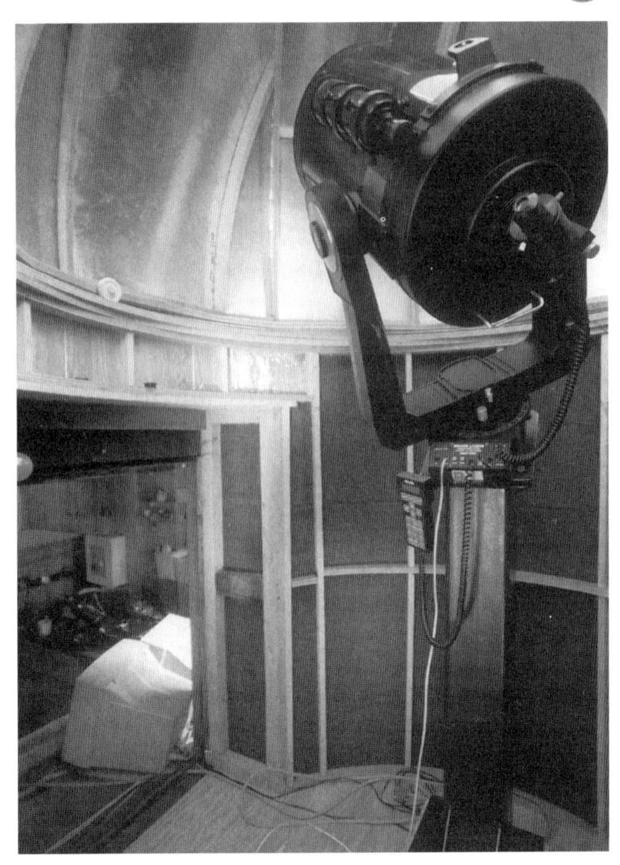

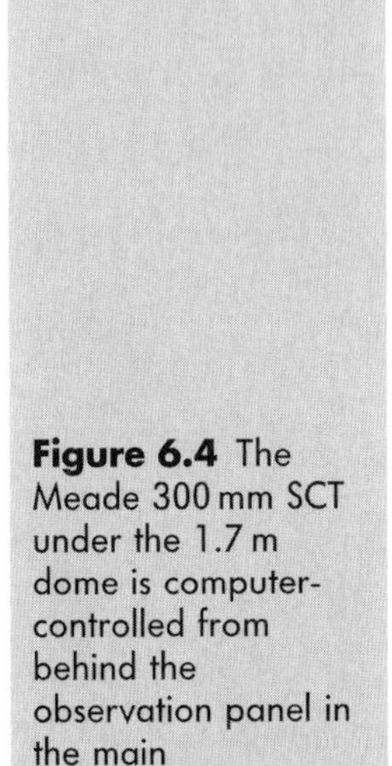

Figure 6.4 The Meade 300 mm SCT under the 1.7 m dome is computer-controlled from behind the observation panel in the main observatory.

With any popular SCT design, the arc described by the eyepiece or camera can be a modest 300 mm (12 in) radius or less, and this makes a small observatory fully practical. When using an eyepiece in a star-diagonal I have found it possible to observe the entire sky (except for the less interesting northern quadrant) from the comfort of a stool or raised chair placed centrally under the dome (see Figure 6.5). Such a feature could be of particular value for a disabled observer.

The cylindrical 1.5 m (5 ft) high timber walls are clad externally in stained softwood boarding on a raised timber platform. Once again to minimise foundation work and disturbing existing paving, the platform is supported off an existing observatory wall and two posts concreted into deep holes. The concrete telescope pier has independent foundations to platform level with a steel column to raise the telescope to the centre of the dome (Figure 6.6).

Figure 6.5 The Meade LX200 SCT under the 1.7 m dome.

The 1.7 m (5 ft 6 in) diameter dome is made of rectangular 450 mm (18 in) wide × 1 mm (1/25 in) thick aluminium sheets curved over a 16-sided plywood frame. In profile the dome appears hemispherical. A piece of flexible garden hose fixed around the shutter aperture forms a tight seal to a single 450 mm (18 in) wide aluminium shutter held in place with clips. The shutter is lifted off in one piece and is set aside when observing. The dome rotates on 75 mm (3 in) diameter nylon wheels running on a plywood track topping the walls.

For CCD observations the telescope is under full computer control and can be operated in comfort from behind a large glass panel in the main observatory (see Figure 6.5). Constructed over a three month period, the materials for the smaller observatory cost about the same as those for a ready-made, prefabricated, rectangular garden shed of similar volume.

Figure 6.6 a The Meade mini-observatory is supported off two posts and the wall of the main observatory. The left-hand window becomes the observation panel separating the two observatories. **b** The Meade observatory takes shape.

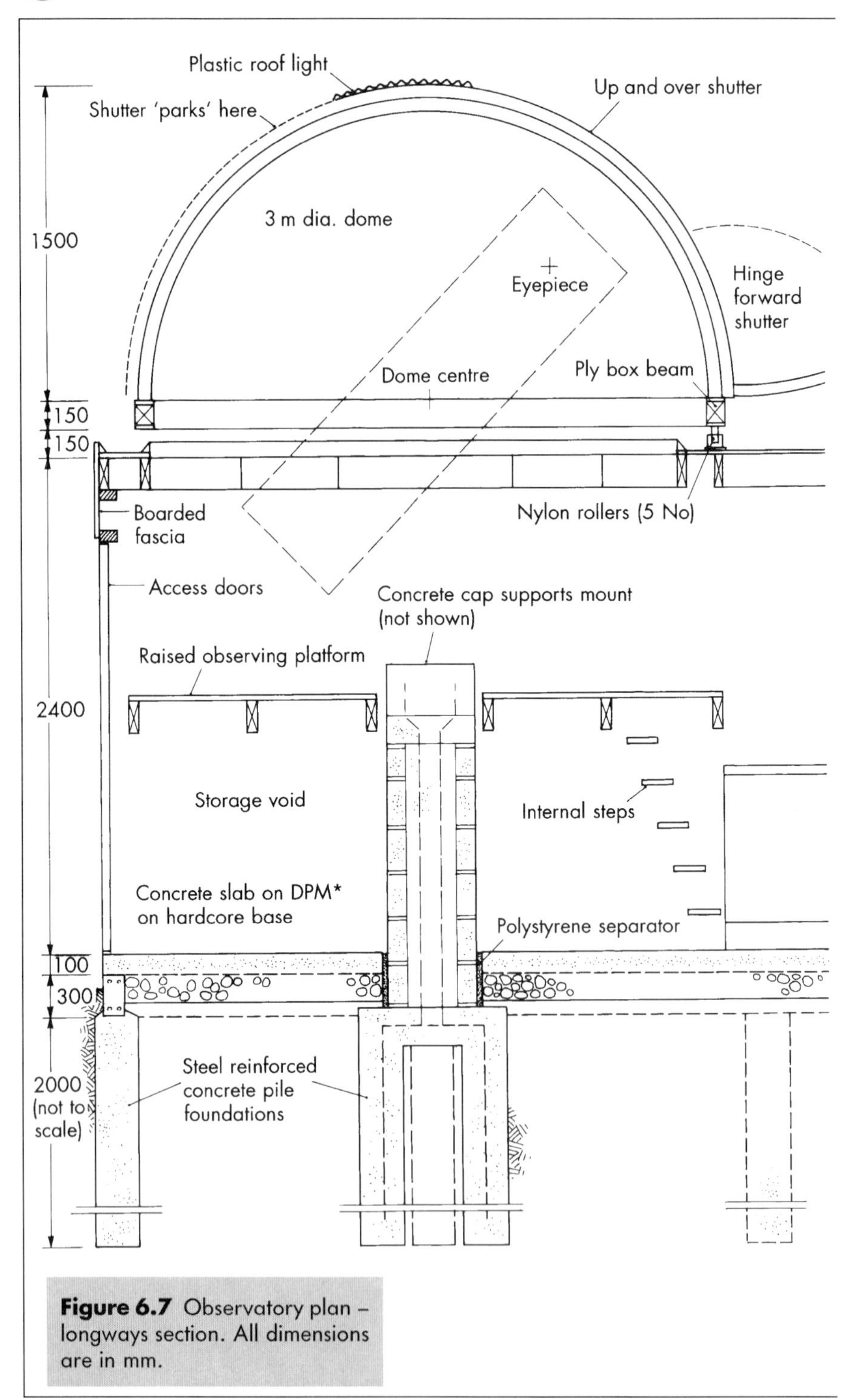

Figure 6.7 Observatory plan – longways section. All dimensions are in mm.

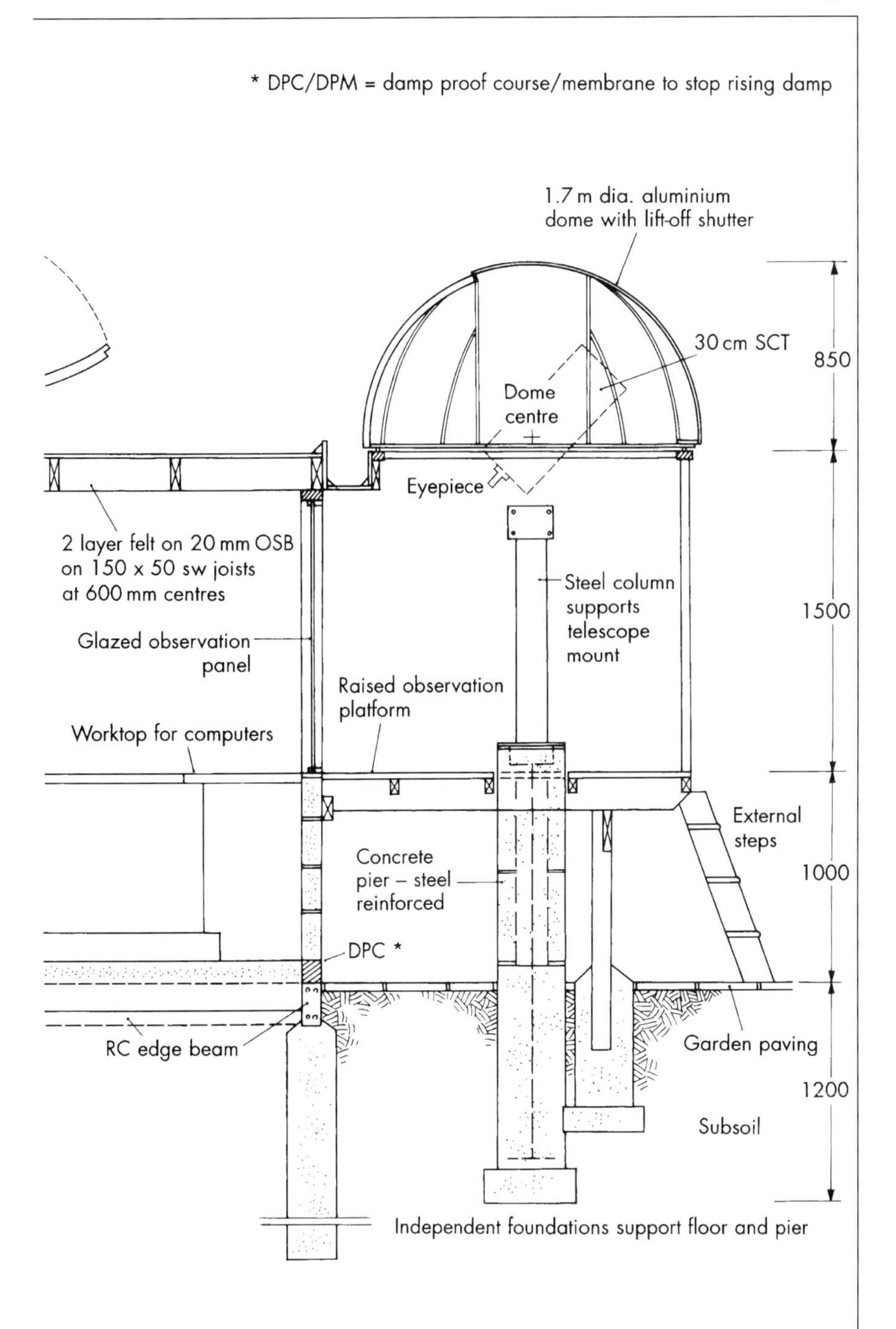
* DPC/DPM = damp proof course/membrane to stop rising damp
1.7 m dia. aluminium dome with lift-off shutter
30 cm SCT
850
Dome centre
Eyepiece
2 layer felt on 20 mm OSB on 150 x 50 sw joists at 600 mm centres
Steel column supports telescope mount
1500
Glazed observation panel
Raised observation platform
Worktop for computers
External steps
Concrete pier – steel reinforced
1000
DPC *
RC edge beam
Garden paving
1200
Subsoil
Independent foundations support floor and pier
© Maurice Gavin RIBA – 1995

In Retrospect

A couple of points caused problems. During the Great Storm of 1987, a puncture in the main flat roof membrane went unnoticed, and eventually caused irredeemable damage to the underlying chipboard roof decking. The whole flat roof was replaced using improved 20 mm ($\frac{3}{4}$ in) OSB boarding and refelted. During this same period my neighbour's garage guttering collapsed; it was inaccessible and regularly saturated part of my south wall. He has subsequently rebuilt his garage clear of my wall with a maintenance gap and the problem has disappeared. A section longways through the observatory is shown in Figure 6.7.

With the exception of steelwork for mounts, purchased optics and SCTs, both observatories and instrumentation were a solo effort, from cutting the first sod to a final lick of paint. This has given me considerable satisfaction. The observatory has proved a complete success and has given much pleasure down the years. Sometimes, I admit, as an occasional retreat from a noisy family! They are now replaced by visiting grandchildren who, under my watchful eye, love the place to spy on the sky.

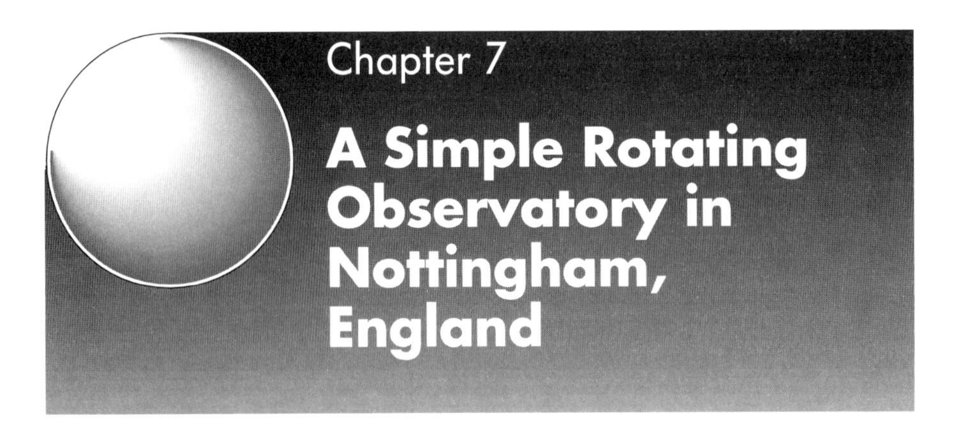

Chapter 7

A Simple Rotating Observatory in Nottingham, England

Alan W. Heath

Figure 7.1 The observatory, showing the roof shutter and extension open.

My observatory, described here, is simple in design, cheap to build and efficient in use.

The Building

The observatory building has a 50 mm × 50 mm (2 in × 2 in) timber frame, 115 mm ($4\frac{1}{2}$ in) wide tongued-and-grooved timbers (floor boards) for the roof shutter and door. The sides and part of the roof are made from asbestos roofing sheets. It is 2.45 m (8 ft) square at the base and 2.6 m (9 ft 6 in) high.

A permanently mounted telescope needs protection from the weather, and living in an urban area calls for screening from local lights. A proper observatory also affords the observer some degree of comfort in the wind and cold of winter.

Domes always present construction problems, so I decided to build the observatory rather like a simple shed, but with a hinged roof-shutter to permit access to the sky.

A hinged extension to the shutter on the same wall as the shutter permits observations at a lower level if required. Even lower objects may be viewed through the open door! The fact that the walls are flat rather than curved is an added convenience that allows for the permanent fixing of charts, photographs, maps etc.

Rotation of the Observatory

The entire structure rotates. It is mounted on an angle-iron ring 2.45 m (8 ft) in diameter which sits on eight equally spaced pulley wheels. The pulley wheels have 12 mm ($\frac{1}{2}$ in) bolts as axles. The ring sits on the wheels (which face upwards).

The angle-iron ring was the only part not made at home. It was necessary to make enquiries at several local engineering firms to find one with suitable facilities to roll the angle-iron into a ring and weld the joint. It is made from 50 mm × 50 mm (2 in × 2 in) angle-iron, flange outwards. It is painted with red metal primer, with a top coat of bitumen paint.

The base of the observatory is a fixture. It is a square frame made from 230 mm × 50 mm (9 in ×

2 in) timber, with angle-pieces made from 75 mm × 50 mm (3 in × 2 in) timber placed across each corner, providing eight points that are of equal distance from the centre. The wheels are placed at these points, and the angle-iron ring lowered on to them. The wheels have to be carefully adjusted before being finally secured, to ensure the ring will turn easily.

The wheels must be level and the bolt which is the axle must be at least half as long again as the wheel is wide in order to allow the wheel to "float" and so compensate for any minor errors in the circularity of the ring itself. The observatory building is secured by screws through holes in the ring.

The wheels and axles are supported by short lengths of angle-iron on each side, the axles passing through holes in these. The whole wheel assembly is further mounted on a piece of sheet steel 150 mm × 150 mm (6 in × 6 in) and then fixed in place on the wooden base. This makes easy any adjustments before finally placing the ring in position.

The wheels are lubricated with a mixture of graphite and grease, which seems to be efficient and fairly quiet – an important point when using the observatory in the small hours of the morning! I haven't noticed any signs of wear and tear in either the wheels or axles since the observatory was put together, quite a few years ago.

The Base

The observatory is on a concrete base which is about 3 m (8 ft 6 in) square, thus providing a slight overlap. The base is approximately 150 mm (6 in) deep. A cubic metre (1.3 cubic yards) of concrete was used, together with some aggregate. In practice it has been found that the observatory can be moved easily by hand, but does not move of its own accord even in a strong wind.

Actually there is no need to have a base unit to support the wheels at all – as I could have fitted the wheels directly to the concrete – but there is the advantage of having a clearance of some 450 mm (1 ft 6 in), so lifting the observatory clear of any snow on the ground without much trouble. Snow on the roof is removed very simply by opening the shutter!

A small louvred ventilator is fitted to one wall to

allow free circulation of air. The temperature does not vary by more than a few degrees from that outside.

Mains electricity enters the building via a conduit through the concrete base, connecting first to a switch box. Electrical safety regulations must be met, and when working outdoors with mains electricity an ELCB (earth leakage circuit breaker) is a valuable safety feature.

All cables for lighting and so on run through the walls, from the junction box near the centre of the roof. From this the cable is fitted with a male-female connector to permit disconnection and periodic removal of "twists" caused by rotating the building more often in one direction than in the other. The observatory was designed to accommodate a 200 mm (8 in) f/8 reflector, something it does easily.

A 300 mm (12 in) reflector later replaced this and this fits as well (see Figure 7.2). As the telescope moves with the roof slit when observing, a "tight fit" is not too much of a problem.

Additional fittings include a drop-leaf table top in one corner for charts, an electric clock showing universal time, a short-wave radio for time signals, and a small cupboard for various accessories including eyepieces, filters and other items.

Figure 7.2 The 300 mm (12 in) Newtonian reflector with the 75 mm (3 in) refractor mounted on the back with a darkened box into which the Sun's image is projected. (This photograph was taken by using a long time exposure and rotating the building so as to use the open door as a moving slit.)

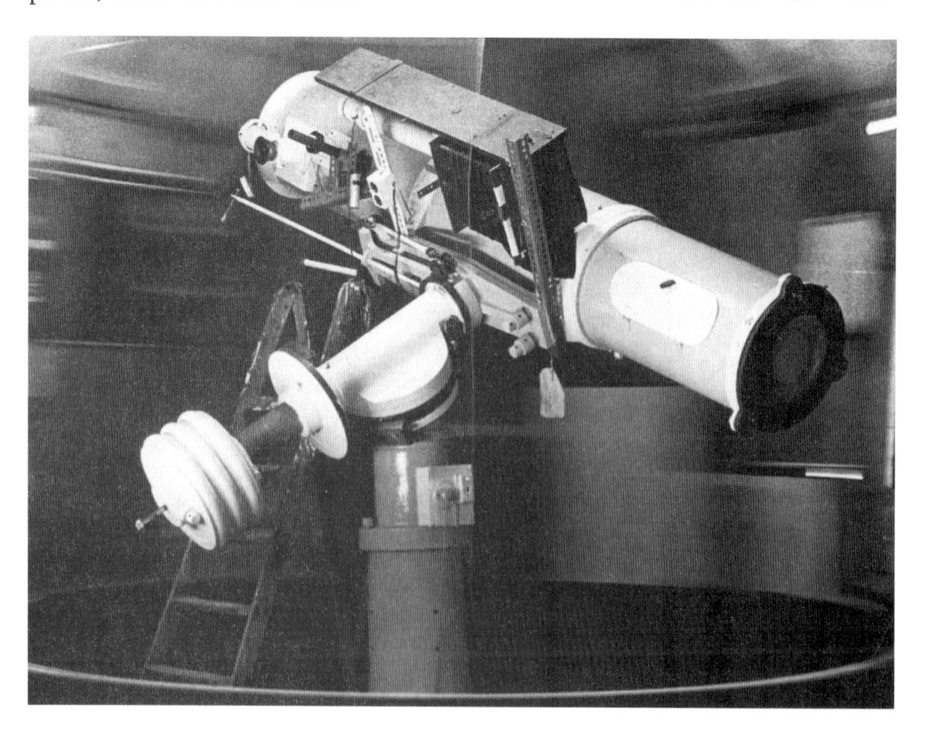

Using the Observatory

The design has proved very efficient. The only maintenance needed is to lubricate the wheels annually and paint the woodwork from time to time. The internal wood frame is creosoted. No replacements have been necessary and it is likely to be good for many more years to come.

I would not change the design much if I were building another observatory. The design can be scaled up or down to suit the size of telescope, but the one described had the advantage that the main covering sheets were a standard size, at least in Britain. Two 8 ft × 4 ft standard sheets cover one side so there was minimum wastage. Today's equivalents are of course 2.5 m × 1.25 mm, which is not so very different.

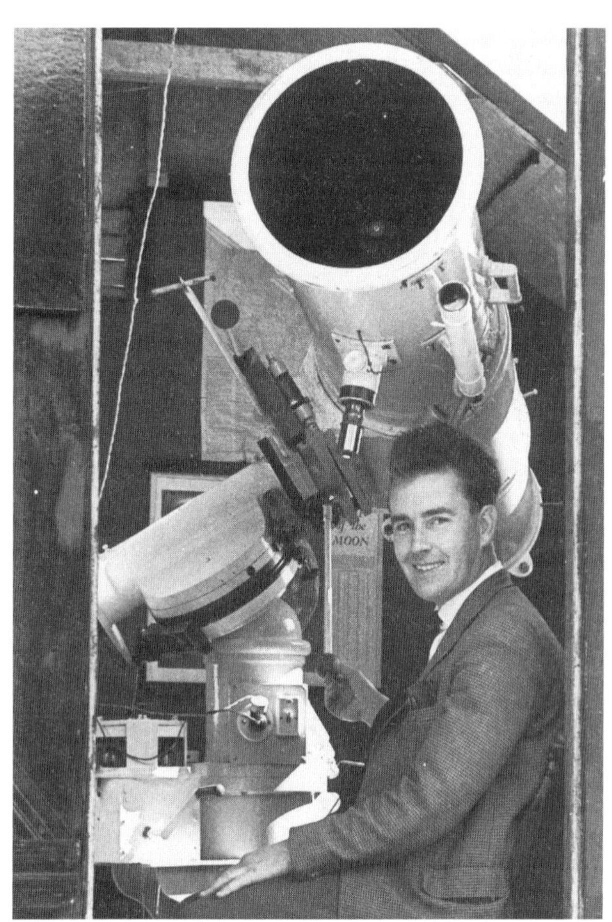

Figure 7.3 The 300 mm (12 in) Newtonian reflector (and the author of this chapter), seen through the main door.

The main telescope is a 300 mm (12 in) Newtonian reflector by Calver (see Figure 7.3), which is the property of the British Astronomical Association and is on loan.

It was originally owned by the Reverend T. E. R. Phillips, a well known planetary observer in the first half of the century. The optics were replaced by H. Wildey in 1961. The telescope is used for lunar and planetary observation, contributing to the various sections of the British Astronomical Association as well as to other organisations overseas. A 75 mm (3 in) Broadhurst Clarkson refractor is mounted on the back of the main telescope and is used for regular observations of the Sun.

Anyone who is remotely handy can construct a building like this, which has the advantage of looking like an ordinary shed and therefore is "neighbour-friendly" while retaining all the benefits of a more sophisticated design.

Chapter 8 The Taunton School Radio Astronomy Observatory in Taunton, England

Trevor Hill

Introduction

Taunton School is a school for boys and girls aged three to eighteen in Somerset in the southwest of England. The school has a reputation for original work in scientific research with records of pupil involvement in scientific endeavours going back seventy-five years.

In June 1988 I decided to involve pupils in astronomy by introducing a certificated course. The course involved practical work and so I built a small glass fibre observatory with an 8 in (200 mm) Newtonian reflecting telescope. It was very difficult to get the pupils at the observatory at a time when the weather was clear and they were not busy.

One day, during another cancelled session when it was raining, a young pupil suggested building a radio telescope because radio waves pass through clouds, and radio sources in the sky can be observed during the day or night. The Sun and the Milky Way emit radio waves that can be detected with simple equipment.

Figure 8.1 Satellite dishes to pick up microwaves at 1420 MHz, and the aerials used for the solar flare detector at 150 MHz. They are photographed on top of the school science block roof.

We built a simple radio telescope to observe the Sun and Milky Way at a radio frequency of 150 megahertz (MHz). In the United Kingdom, local radio stations broadcast between 88 and 108 MHz in the very high frequency (VHF) band and so we wanted to look at radio waves from space at a slightly higher frequency than this.

In March 1989 we picked up radio waves from a huge solar flare. It was very exciting. The Northern Lights were visible that evening in the sky, even from Somerset. This was very rewarding and so the group decided to build even bigger and better radio telescopes. This was how the Taunton School Radio Observatory was established. There are now four radio telescopes in operation. They pick up radio waves from space at four different frequencies. The best observations are made when solar flares occur. We can observe radio waves from other objects too. A radio galaxy in Cygnus and a supernova remnant, the remains of a dead star in the constellation of Cassiopeia, have been observed.

It is good to give young people in a school the chance to operate a radio telescope and study the results – this is the best way to learn science. Our pupils often leave school to carry on an interest in astronomy, and many have gone into careers in science and engineering.

The VHF Radio Telescope – A Solar Flare Detector

To understand a radio telescope, I start by explaining a television system. To pick up the television radio waves, an aerial is needed. The radio waves fall upon the aerial, and small electrical signals are induced in it. These can be made larger by using a booster amplifier connected to the aerial. These larger signals then pass down coaxial cable into the back of the television set, where a tuned radio receiver picks up the signals. The signals are then used to produce the picture on the television screen. Ground-based television (in the UK) works by picking up ultra-high frequency (UHF) radio waves; satellite television needs dishes that pick up microwaves from satellites orbiting the earth.

We build radio telescope systems like the television system. An aerial or dish is needed to pick up radio waves, not from an Earth-based transmitter or satellite like the television does, but from distant astronomical radio sources like dead star remains and active radio galaxies.

Some of these radio waves come from sources that are thousands of millions of light years distant. When they fall upon our aerials and dishes, the electrical signals that they produce are very tiny. These must be amplified. As with a satellite TV receiver, a pre-amplifier is fitted close to the aerial to provide low-noise amplification of the signal, which then passes down a coaxial cable to the radio receivers, tuned to the frequency that we are interested in. We tune the receiver to a very faint hiss where there is no station – anyone can hear this by just taking any radio and tuning it away from any man-made radio station.

In our radio receivers, the hiss produces a tiny voltage that is accurately measured and plotted on a computer screen. When we point the aerial at the Sun, the hiss increases to a roaring sound and the computer screen shows a rise in signal strength.

If a solar flare occurs when the sun is above the horizon, a huge increase in signal strength can be seen on the computer screen. Our aerials do not track objects like an optical telescope; they are fixed relative to the ground. Rotation of the Earth causes radio sources to drift past, allowing us to observe astronomical radio sources.

A diagram of this simple radio telescope is given in Figure 8.2.

A picture of the aerials used for this system can be seen in Figure 8.3. The aerials pick up radio waves at a frequency of 150 MHz (in the VHF band). We observe radio waves from the sun, and particularly radio waves from solar flares.

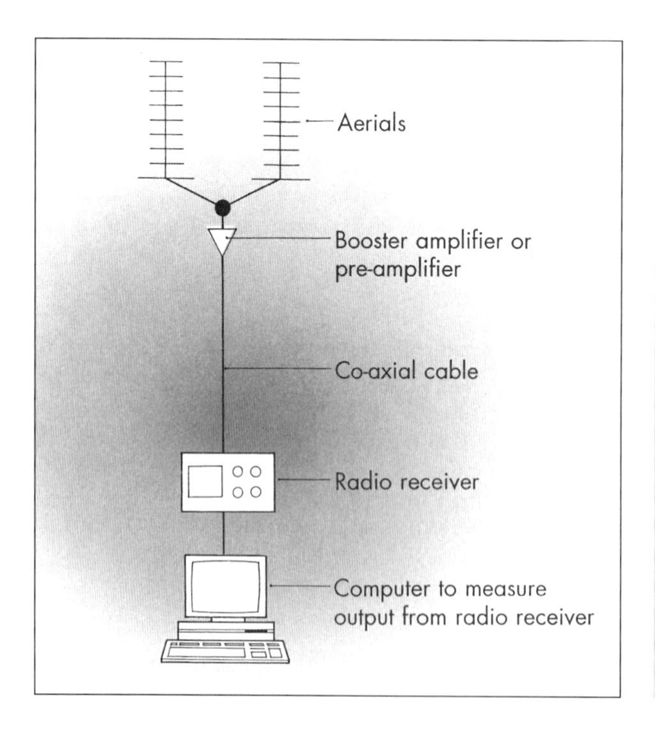

Figure 8.2 A radio telescope (150 MHz VHF) used as a solar flare detector.

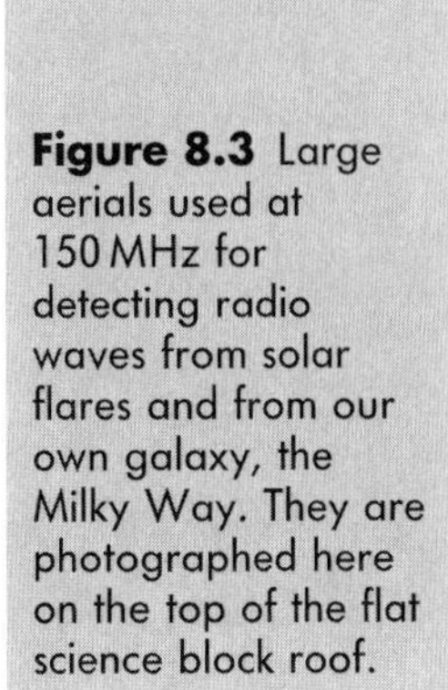

Figure 8.3 Large aerials used at 150 MHz for detecting radio waves from solar flares and from our own galaxy, the Milky Way. They are photographed here on the top of the flat science block roof.

The system has also been used to pick up VHF radio waves from our own Galaxy, the Milky Way. We can detect a solar flare (even on cloudy days!) and it does not actually matter where the aerial is pointing, because the radio waves are so strong. It can be used as a solar flare detector, allowing us to predict when the Northern Lights may be visible.

The aerials can be bought from any retailer that sells aerials for amateur radio enthusiasts. Just ask for a Yagi-type aerial for operation at 150 MHz. Television aerials are Yagi-aerials designed (in the UK) for operating at about 600 MHz, which is in the UHF band.

Our Yagi aerials are just large TV aerials – they are about four times bigger and have only ten elements (whereas UHF TV aerials have about twenty-five or more elements). A suitable preamplifier can also be purchased from a similar retailer – preamplifiers are

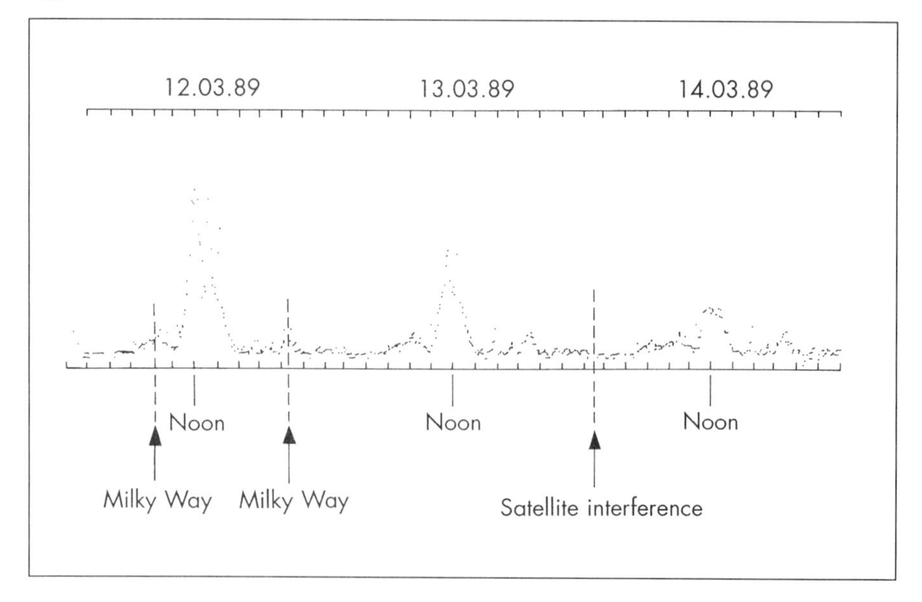

Figure 8.4 Radio waves picked up from the sun at 150 MHz during March 1989, using the solar flare detector system shown in Figure 8.2.

quite cheap. The coaxial cable is the kind used for carrying television signals.

Our main radio receiver was bought second-hand. The whole system cost about £200 (1995) to make, excluding the computer. Almost any PC can be used with a suitable interface – an old Acorn BBC computer is ideal. A chart recorder could be used instead and would be just as good.

An example of the results using this system can be seen in Figure 8.4. This shows the radio signals received at 150MHz over three days, when a large sunspot was present that was producing solar flares.

The Microwave Radio Telescope

We have made several different kinds of radio telescope since the solar flare detector just described. Our most recent system uses two large satellite dishes that pick up microwaves from space at a frequency of 1420 MHz. The two large dishes and the aerials for the solar flare detector can be seen in Figure 8.1.

The dishes were purchased and sent across from the USA. The pupils assembled them in their spare time – they took three months to build.

The dishes are 4 m (13 ft) in diameter, and can move in altitude to point at different angles. A hydraulic ram pushes up a metal ramp that the dish is fixed to. The dishes do not move in azimuth (left and right); only up and down. Rotation of the Earth scans objects as they drift past. The dishes are very directional and must be pointed quite accurately. Pointing at the Sun is easy if the Sun is shining because we can use shadows, but on cloudy days or when we point it at objects other than the sun, there are a protractor and a plumb line that allow the dishes to be set at the required angle.

Our control room is shown in Figure 8.5. The control room is indoors, warm and dry, and contains all the radio receivers and computers needed to process the microwave signals.

This microwave radio telescope uses two dishes. The signals from each are added together before they are sent to the control room, forming a special kind of radio telescope system known as a radio interferometer. It is much more sensitive than the solar flare detector and can detect very weak radio signals from space. We have used it to pick up radio waves from distant quasars, the Andromeda Galaxy, the Crab Nebula, and many different kinds of radio galaxy.

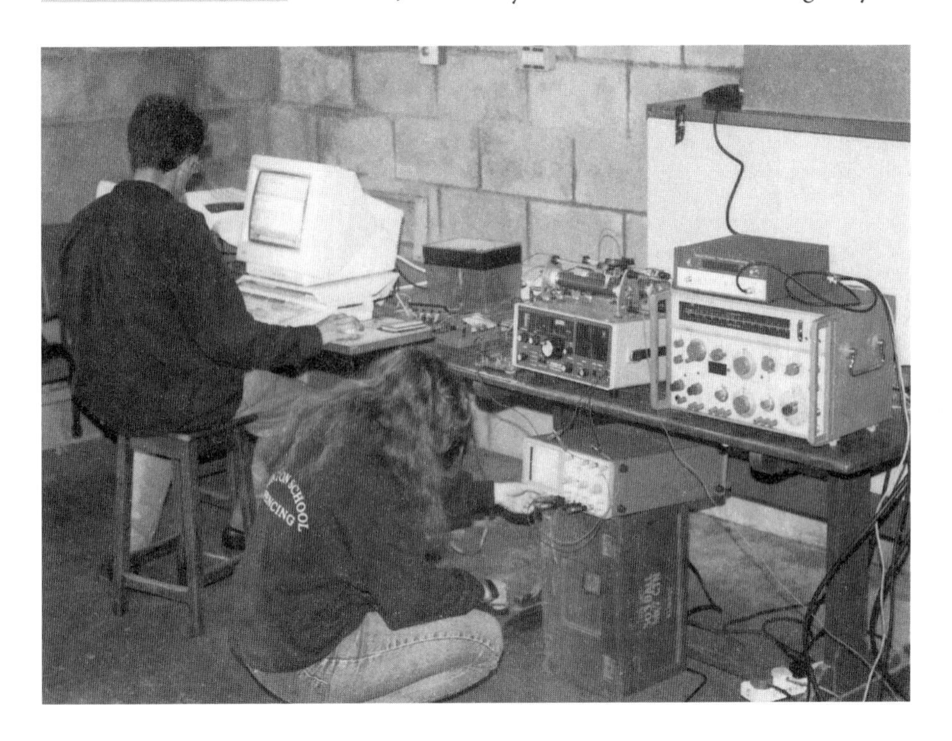

Figure 8.5 The radio observatory control room contains the radio receivers. The computers measure the signals produced by the radio receivers that are connected to the aerials on the science block roof. There are four radio telescopes operating, but only one computer which measures all the signals. When a solar flare occurs, the room is a buzz of activity!

A Radio Telescope for Observing Jupiter

The planet Jupiter emits radio waves that are very strong around 20 MHz, which is in the high frequency (HF) band. In Figure 8.6, there is a picture of the aerial system built by our pupils to observe these radio waves. It is desirable to build these aerials as high off the ground as possible, to get a better radio "view" of Jupiter. We built two scaffold towers so that we could use two aerials and observe Jupiter using an interferometer at 20 MHz.

In July 1994, we observed radio waves produced from Jupiter when Comet Shoemaker-Levy-9 collided with it. The Taunton School Radio Astronomy Observatory was among the first in the world to pick up radio waves during the first impact of the fragment A. We also observed radio waves emitted during the collisions of fragments H and Q, which shows that useful work can be carried out by amateurs using simple inexpensive equipment.

Figure 8.6 One of the two scaffold towers built to observe Jupiter's radio waves.

Conclusion

Our radio observatory contains four different radio telescopes and can make useful radio observations of the sky. There is a lot that can be learned about aerials, electronics, radio and computing as well as astronomy and physics. Radio astronomy is exciting and radio observatories will continue to make observations and achieve results leading to a greater understanding of our universe.

Young people learn quickly when they have the opportunity to actually use equipment in an operational observatory. It is often the best way to learn science. Remember, the pupil of today is the research scientist of tomorrow!

Chapter 9

The Starlight CCD Observatory in Binfield, England

Terry Platt

History of the Observatory

I have been an active amateur astronomer since the age of eleven, when an old encyclopedia inspired me to look at the sky and comet Arend-Roland gave me something special to look at! My first telescope was a 32 mm ($1\frac{1}{4}$ in) refractor and no observatory was required, but a rapid succession of ever-larger instruments began to make the idea of an enclosure much more attractive. However, money and time prevented anything permanent from being built until many years after my first look at that bright comet in 1957.

In 1984 my wife and I purchased a bungalow in the outskirts of Binfield, near Bracknell in Berkshire. This had a reasonably large back garden and a concrete plinth where a garage had once stood, close to the north fence. The location was fairly dark and the view to the south was quite clear of obstructions, apart from a few small trees in the neighbour's garden. After spending a (minimal!) time getting the new home into some kind of order, I began to plan the design of a workshop and observatory to occupy the ready-made base.

Figure 9.1 Terry Platt's observatory and workshop from the south-west.

Planning the Construction

I wanted to combine the observatory with a workshop for electronics work, and mirror grinding and other such activities, as I enjoy making my own telescopes and devices such as CCD cameras. So I decided to build a two-storey structure with the lower section equipped with workbenches and a lathe. The 2.8 m × 4 m (9 ft × 13 ft) concrete base offered a reasonable area for a workshop and it would be possible to add a dome of some kind for an observatory capable of enclosing my 300 mm (12 in) f/5 Newtonian reflector, which was my telescope at that time.

Many amateurs have used metal, glass fibre and concrete for observatory construction, and these are very suitable materials. However, I like to work with wood and it does have the advantage of easy availability and being less prone to condensation problems, although it also needs regular maintenance. A secondary but important consideration was the aesthetic appearance of the structure, which needed to merge into the surroundings as well as possible in this green-belt area.

A basic design was sketched, in which the outer shell would be a shiplap timber skin over a frame of 75 mm × 50 mm (3 in × 2 in) treated pine beams. The overall shape would be a 2.8 m × 4 m (9 ft × 13 ft) rectangle with the western end roofed with a 2.8 m (9 ft) plywood dome, and the remaining eastern portion covered by a roof with a skylight. An upper floor at about 2.2 m (7 ft 3 in) height would cover the area under the dome, with access via a stairway on the north side. This construction would place the dome base-ring at about 2.7 m (9 ft) above ground level and considerably improve the field of view of the telescope, which would be badly restricted by local trees and our house if mounted at ground level.

These initial thoughts were all based on the assumption that the dome diameter had to clear a relatively short-focus Newtonian reflector and so would not need to be greater than about 2.8 m (9 ft). However, although a dome of this size was built and fitted during 1984, I later replaced it with a 3.9 m (12 ft 10 in) construction, which was necessary to clear my new Buchroeder Tri-Schiefspiegler, installed in 1988. This later dome incorporates improvements on the original and so will be the one described in the remainder of this chapter.

Building the Support Structure

The frame of the workshop section was to be the first part to be built and a source of good quality, inexpensive timber was needed. 75 mm × 50 mm (3 in × 2 in) beams are easily obtained at almost any woodyard, but the price can be quite high, especially if you want the timber pretreated against rot. I decided to see if the wood from a demolition contractor's yard would be satisfactory, and was pleased to discover that much of it was rot-protected and that the price would be less than half that of new wood!

The two long sides of the frame were built horizontally on our back lawn by using a combination of half-lap joints and galvanised 75 mm (3 in) nails. The desired outline was laid out in pre-cut wood sections and a grid of horizontal and vertical cross-members added at stress points and to outline windows and the door etc. This job took only a couple of days to complete and then the rest of the family were dragooned into helping to push the frames into an upright position on the north and south sides of the concrete base. The west and east ends were then added by cutting and nailing the remaining frame members into place with the whole structure in its final location on the base. Once the frame was completed, windows made from Polycell secondary double glazing channel with 4 mm glass were added on the east, south and west sides and then the remainder of the external surface was panelled with 100 mm × 18 mm (4 in × $\frac{3}{4}$ in) shiplap wooden planking. This basic structure has remained unchanged since 1984, but many other details have evolved up to the present day. The remainder of this description is of the observatory as it exists currently in 1995.

As the lower room is used as a workshop, extra horizontal members of 100 mm × 75 mm (4 in × 3 in) cross-section are fitted to form the supports for a workbench at each end of the structure, and these are planked over with 150 mm × 32 mm (6 in × 1$\frac{1}{4}$ in) floorboards to give a pair of 2.8 m × 0.5 m (9 ft × 1 ft 10 in) work surfaces. One of these is used for metal and woodworking, while the other provides a work surface for electronic design and assembly.

Access to the observatory level is provided by adding a simple wooden stairway to a trap door on the north side of the workshop. This is made from sections of 200 mm × 25 mm (8 in × 1 in) wooden planking from an old packing case, with 50 mm (2 in) square tread bearers to reinforce the step supports. The observatory floor itself is supported on four 100 mm × 50 mm (4 in × 2 in) beams and is made from 150 mm × 32 mm (6 in × 1$\frac{1}{4}$) floorboards, as per the bench surfaces.

Because the base of the telescope mount needs to be approximately 1.4 m (12 ft 7 in) off the floor, it is supported on a concrete block column, built onto the old concrete garage plinth. This is not a recommended method of building a telescope pier, as it is fairly prone to surface wave vibration from nearby roads etc., but this has proved to be only a minor problem in my fairly quiet location. The column is composed of twelve pairs of 460 mm × 230 mm × 230 mm (18 in × 9 in × 9 in) concrete blocks, cemented together to form a 460 mm (18 in) square pillar,

1.4 m (4 ft 6 in) tall. To give added strength, 18 mm ($\frac{3}{4}$ in) studding was fitted through holes in the topmost blocks and tightened into place with nuts and washers. This prevents the top end of the pillar from being split apart by stress from the steel mounting base-plate, which is held in place by 12 mm ($\frac{3}{4}$ in) expanding bolts. The column was positioned some distance to the south of the centre point of the observatory, so that the fork mount of the telescope would overhang into the exact middle of the dome and minimise the dome diameter needed to clear the telescope.

The mounting itself is a hybrid assembly of steel angle, aluminium and wood, which was originally assembled with a 400 mm (15$\frac{3}{4}$ in) diameter worm and wheel drive on a 100 mm (4 in) diameter tubular steel shaft (see Figure 9.2). This drive system was never sufficiently stable for long-exposure work and it was replaced with a friction roller system (similar to that used by Ron Arbour) during 1993. It now consists of a 760 mm (2ft 5 in) diameter × 6 mm ($\frac{1}{4}$ in) thick hard aluminium disc, driven by a spring-loaded 25 mm (1 in) stainless steel roller (see Figure 9.3). The roller is in turn driven by a 200 steps per rev stepping motor running in half-step mode via a 1250:1 gearbox. A crystal-controlled programmable divider supplies the motor with 176 Hz for sidereal driving and can be programmed in 0.001 Hz steps to set any precise rate that may be needed. This combination gives a good performance on unguided exposures of up to 10 minutes, using an image scale of 3 seconds of

Figure 9.2 The declination tangent arm drive and stepper motor.

Figure 9.3 The friction roller RA drive.

arc per pixel on a deep-sky CCD camera, and will keep planets within a 20 seconds of arc error box for over one hour without correction.

The polar axle is supported by a 150 mm (6 in) diameter ballrace mounted in an aluminium cell on the 18 mm ($\frac{3}{4}$ in) thick steel polar plate of the base frame, the whole assembly being carried on a 450 mm (1 ft 6 in) square base-plate with 50 mm × 50 mm (2 in × 2 in) steel angle struts. This strong frame is needed to carry the overhanging weight of the two 320 mm (12$\frac{1}{2}$ in) telescopes on a fork mount, which are the current occupants of the observatory. The fork is built up from 18 mm ($\frac{3}{4}$ in) exterior quality plywood and is designed to be both strong and light. A rigid base-plate consists of three layers of ply, glued and stacked together, with an overall size of 950 mm × 475 mm (3 ft 2 in × 1 ft 7 in). The fork tines are single pieces of 18 mm ($\frac{3}{4}$ in) ply, 950 mm (3 ft 2 in) long and tapering from 475 mm (1 ft 7 in) wide at the base to 200 mm (8 in) at the top end. They are set 650 mm (2 ft 2 in) apart which leaves about 130 mm (5 in) between the outer surfaces and the ends of the fork base, a space which is occupied by ply reinforcing ribs on either side of the declination axis. Four 100 mm (4 in) long × 12 mm ($\frac{1}{2}$ in) diameter bolts screw the fork base down onto the friction drive disc and polar axis end-plate. I find that this fork assembly is exceptionally rigid and I can stand on the ends of the tines without any obvious deflection occurring, a test which many metal forks would fail.

Two short lengths of slotted steel angle are attached to the fork tine tips and these carry pillow block ballrace assemblies. Two lengths of 15 mm ($\frac{5}{8}$ in) diameter steel studding pass through the ballraces and are fixed with washers and lock nuts to the walls of the main telescope tube assembly to form the declination axis. Declination control is provided by a "tangent arm" drive, which consists of a 900 mm × 90 mm × 6 mm (35 in × $3\frac{1}{2}$ in × $\frac{1}{4}$ in) thick aluminium arm with friction blocks acting on a 200 mm (8 in) diameter aluminium disc attached to the telescope side wall. The lower end of this arm engages with a travelling nut on a stepper-motor-driven threaded rod, giving a total angular travel of about $\pm 7°$. This arrangement has been very satisfactory, with virtually no backlash and precise pointing capability. However, it does not lend itself to fully automatic operation and may be replaced with a 360° drive system at some future time.

The Telescopes

My observatory is designed for use with CCD imagers and the telescopes are not normally used visually for extended periods. Because of this bias towards electronic imaging, the telescopes are somewhat unusual and one of them is not capable of visual use at all. My main interest is planetary observation, and so the primary instrument is a long-focus, off-axis reflector based on the "Schiefspiegler" design, pioneered by Anton Kutter and refined by Dick Buchroeder. This first appeared in the 1950s as a tilted two-mirror arrangement with a long-focus concave primary and convex secondary, giving "refractor" definition in apertures of up to about 125 mm (5 in). The two-mirror design had too much residual aberration to be used at large apertures, but when a third tilted mirror was added by Buchroeder, the maximum aperture could be increased to about 320 mm ($12\frac{1}{2}$ in) before the aberration was unacceptable. This design appeared in *Telescope Making* magazine number 28 (Fall 1986), and I decided that it would be an ideal instrument for planetary CCD work.

The primary mirror of my Schiefspiegler is made from a 318 mm × 60 mm ($12\frac{1}{2}$ in × $2\frac{1}{2}$ in) Pyrex blank, which I ground and polished to a f/12 elliptical figure (approx. 50% of a paraboloid). The secondary is made from a 150 mm × 20 mm (6 in × $\frac{3}{4}$ in) plate glass blank, ground and polished to a spherical convex curve with the same radius as the primary (it is actually the central part of the tool disk used to make the primary mirror), and the tertiary mirror is from a similar blank, but polished to a very weak spherical concave curve with a radius of 53,238 mm! Because three reflections would give a laterally inverted image and also lead to a rather awkward eyepiece location, an extra small flat mirror is added to send the beam out of the tube at 45° just above the main mirror cell. The complete system has a focal ratio of f/20 and gives excellent planetary images, although it is badly compromised by the British seeing conditions! The complete optical layout is shown below (Figure 9.4).

Because of the long focal length of the Schiefspiegler, a separate instrument is needed for deep-sky work. For some time this was provided by a 200 mm (8 in) f/5 Newtonian reflector (originally made to view comet Kohoutek in 1973!) which piggybacked on the Schiefspiegler tube assembly. This provided some good images of nebulae and galaxies, but faint objects needed exposure times which were rather too long for

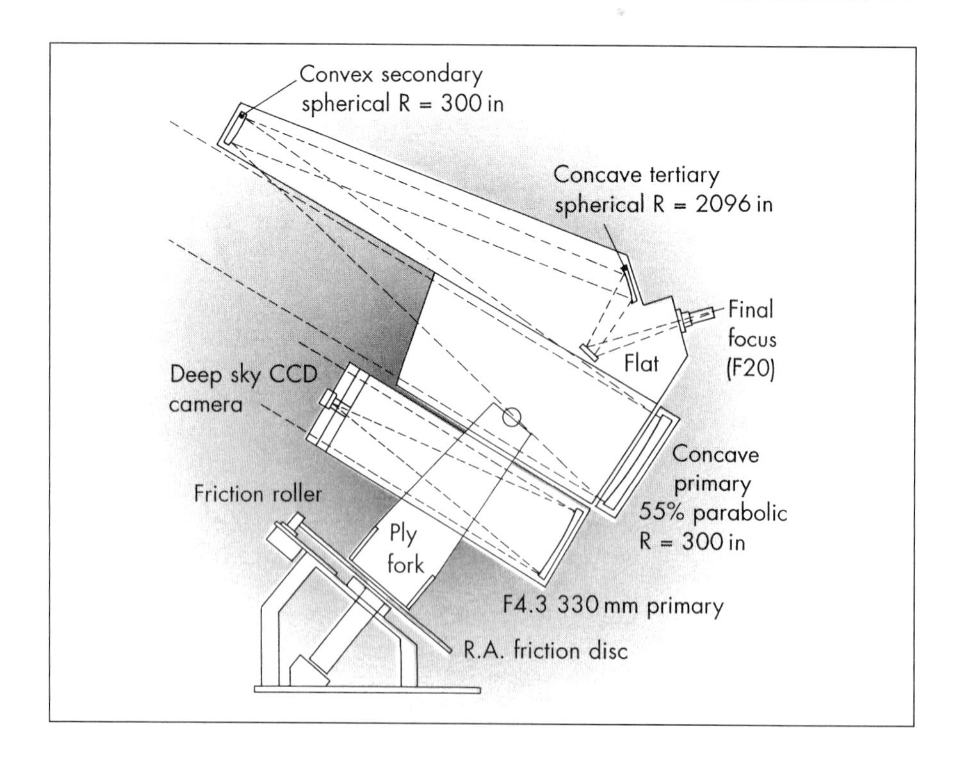

Figure 9.4 Quad-schiefspiegler and deep-sky camera.

the tracking ability of the drive system and the image scale was just a little too small for most galaxies. The result was that during early 1995 I decided to make a dedicated camera system, using a spare 330 mm (13 in) plate glass blank and the focus assembly from an old telephoto lens. The blank was ground and polished to an f/4.1 paraboloidal curve and mounted in a simple square plywood tube, slung below the Tri-Schiefspiegler so as to keep the centre of gravity as low as possible. Some geometry indicated that the camera should have an unobstructed view down to about 10° from the southern horizon and I decided that this would be adequate, considering the light pollution and haze at this level. A strong four-legged spider was made from 3 mm ($\frac{1}{8}$ in) thick aluminium strip and the telephoto lens barrel (minus lenses) mounted at its centre. The lens barrel has a Pentax-Praktica 42 mm × 1 mm thread at its upper end and this allows me to screw on a CCD camera so that the CCD chip lies close to the focal plane of the primary mirror. The focus screw of the lens barrel then allows the camera to be precisely positioned at the prime focus of the mirror, with enough adjustment range to permit various different cameras to be substituted, if required. The images provided by this optical assembly are, of course, laterally inverted, but the CCD has a double-ended readout register, which allows it to be downloaded in a laterally reversed mode to compensate. Alternatively, the images are readily reversed when the computer is processing the results.

The combination of telescopes described above is well suited to my requirements and is unlikely to be changed in the near future, although "aperture fever" can strike at any time!

The Dome

The component that causes the greatest headache for most observatory designers is the "lid". There are so many possibilities, each with its own drawbacks, and the temptation is to start without having considered all the problems that might arise. I am no exception to this rule and my first attempt at a rotating observatory dome resulted in something with insufficient space and an awkward door on the observing slit. This first dome was good enough to use for several years, but eventually became such a liability that it just had to be replaced with something larger and more "user-friendly". Many hours of sketching and calculation went into the design of the new dome, with the result that it is far more convenient to use, and did not cost a great deal to make. Of course, nothing is ever perfect, and I will probably modify some details in the near future, but I am quite happy with the overall result.

As with all the other parts of the observatory, the dome is constructed from wood. This is easy to work, readily available, and quite weather-resistant if treated with modern coatings. Also, the overall weight is not as great as would be the case if metal or an adequate thickness of fibreglass had been used. The first dome had been a true hemisphere and this had resulted in a lot of complicated cutting and shaping, along with the need to use large numbers of nails to maintain the distorted shape of the plywood gores. The new dome avoids this by being polygonal and so is composed of about thirty-three trapezoidal flat panels, attached to a wooden frame by brass screws, a very much simpler shape to make!

Another important decision that had to be made was how the dome could be made to rotate without a great deal of force being required. The first dome had been supported on eight small plastic "buggy" wheels from the local hardware shop, but these did not provide enough support to prevent the dome wall from sagging between the wheels, and would not be strong enough to carry a greater weight.

I eventually decided on a distributed support composed of a large number of golfballs, which I naively expected to roll around a simple channel without bunching and jamming – big mistake number one! The balls would be retained on a wooden platform by arcs of 18 mm ($\frac{3}{4}$ in) ply and would carry the base-ring of the dome in a similar manner to the operation of a ballrace. This idea was eventually put into action, but as you will see, it was considerably modified in the final version.

The overall dome diameter that was required to clear the Schiefspiegler was about 3.8 m ($12\frac{1}{2}$ ft) and so was substantially larger than the 2.75 m square observatory floor. Fortunately, 3.87 m (12 ft 9 in) is the diagonal width of the floor, and so a 3.87 metre dome is just large enough to clear the support structure at all points. This seemed to be an ideal solution and so a set of 50 mm × 100 mm (2 in × 4 in) support beams was fitted to the top of the observatory walls to form a second square frame at 45° degrees to the main building outline. The eight-pointed, star-shaped grid of beams provided sufficient support to carry an extended circular platform of the required 3.87 m (12 ft 9 in) outside diameter, composed of 18 mm ($\frac{3}{4}$ in) exterior-quality plywood. The inside diameter of this platform was cut to match the original floor width of 2.75 m (9 ft) and so provided both a continuous support for the dome bearing track and a broad and very useful storage shelf for eyepieces, star charts and other accessories. Each of the quadrants required to make up the circular platform is cut from a single 2.44 m × 1.22 m (8 ft × 4 ft) sheet of spruce ply, marked out with a wooden radius arm and pencil while laid out on the patio. They were cut to shape with a jig saw and fixed in place with large zinc-plated screws.

The next operation was to add a circular track for the golfball-bearings and this was done by cutting arcs of 18 mm ($\frac{3}{4}$ in) ply, approximately 50 mm (2 in) wide and of radii that matched the desired inner and outer track diameters. The facing edges of these arcs were bevelled so as to provide a broad restraining surface for the golfballs and to avoid abrasion of their outer skins, and then they were screwed into place on the periphery of the platform. The result is a circular channel about 45 mm ($1\frac{3}{4}$ in) wide and 75 mm (3 in) from the edge of the platform, in which standard 43 mm golfballs will freely roll.

Having built a base for the observatory dome, construction of the dome framework was begun on the basis of drawings which I had generated on my PC, using the drafting package "FastCAD" (see Figure 9.5). The most important design parameter was to be able to cut the various parts from standard 2.44 m × 1.22 m (8 ft × 4 ft) sheets of plywood, with a minimum of wastage. This was possible for most of the ribs and panels, but a few had to be assembled from two pieces, locked together with simple dovetail joints and PVA wood glue. These large parts were mainly for use in the door assembly, but two were needed to provide the sides of the slit opening. As I mentioned above, I avoided curved sections in the skinning of this dome by building up an approximate hemisphere from polygonal panels. The outer surfaces of the dome ribs are therefore shaped in three straight cuts,

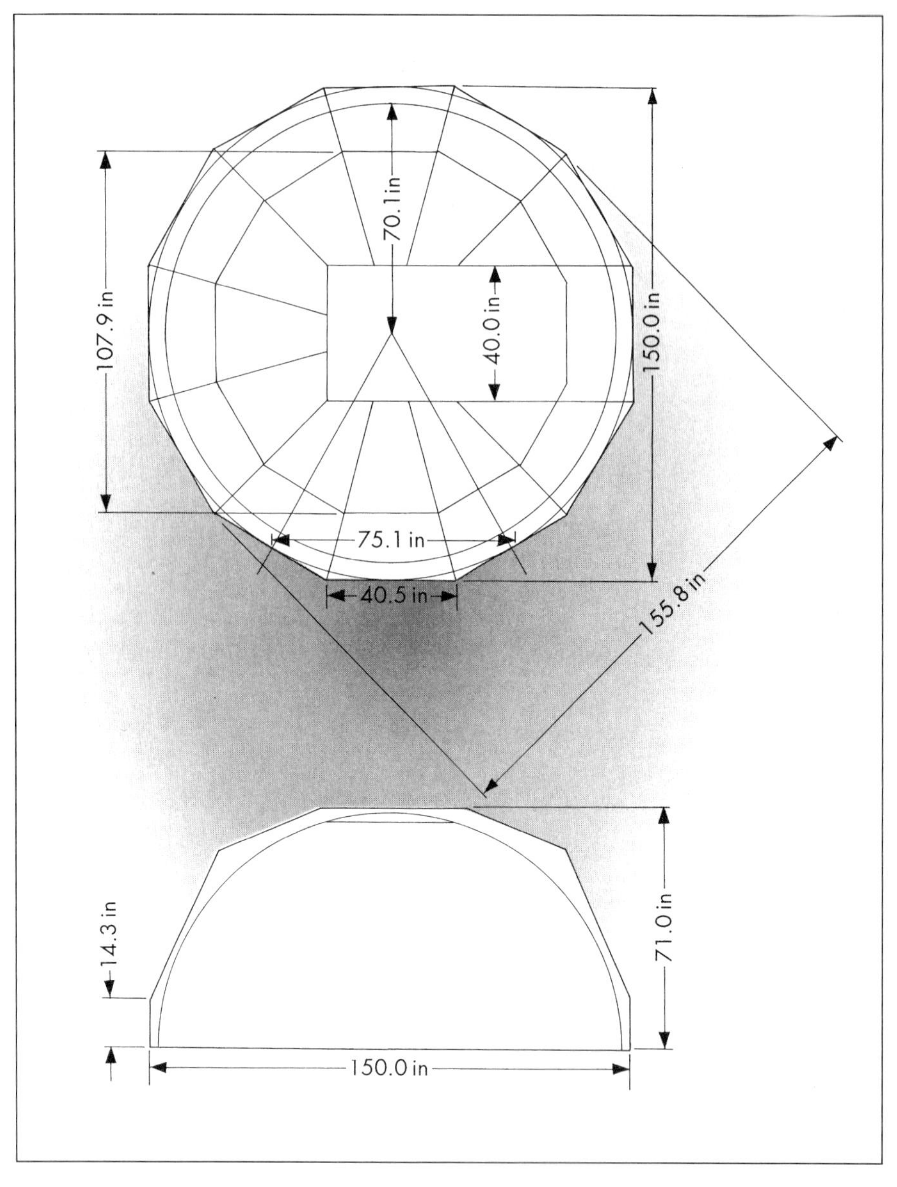

forming tangents to the ideal circular shape, while the inside surfaces are simple arcs with a radius of about 1.8 m (6 ft). All are cut from 18 mm ($\frac{3}{4}$ in) exterior plywood and bevelled on the outer edges to match the angles of the skin panels. The first stage in assembling the frame was to lay out the circular base-ring, which also acts as the ballrace rotor by resting on the ring of golfballs. This is composed of six arcs of 18 mm ($\frac{3}{4}$ in) ply, 130 mm ($5\frac{1}{4}$ in) wide and joined

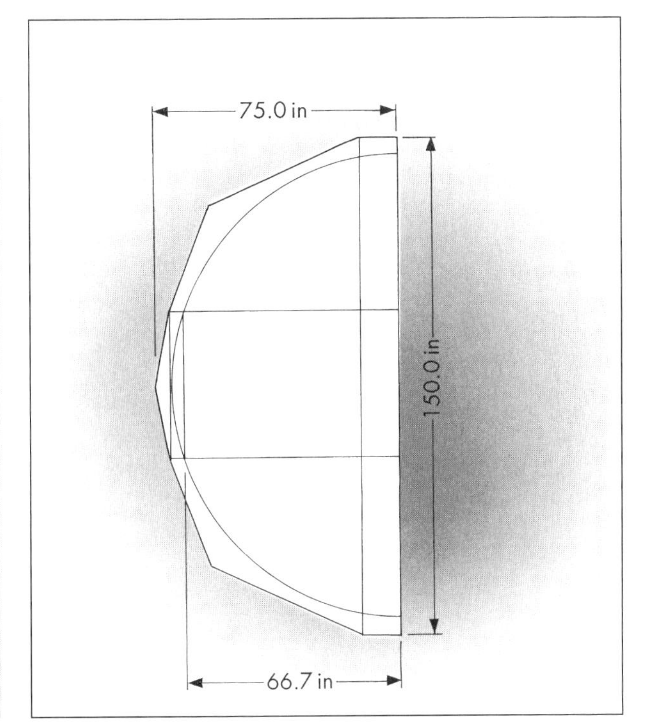

Figure 9.5 (*Opposite page and right*) General assembly diagram for the observatory dome.

together by interlocking dovetail couplings. The joints are reinforced by short bridging sections of ply screwed to the top side of the arcs, and a circumferential ring of 25 mm (1 in) square section pine is screwed to the underside of the ring, to form a restraining "collar" and so keep the ring centred on the balltrack.

Once the base-ring was completed, the side walls and transom for the observing slit were attached to the ring, using two-part plastic joints (as sold by all hardware shops) and supported in place by fitting an extra dome rib directly opposite to the slit aperture. Using the two-part joints gives the builder the ability to remove and replace various parts of the assembly at will, and simplifies the process of "tweaking" the various ribs to get a true and symmetrical framework, a difficult problem at the best of times. The next stage was to fit each of the other ribs, taking care not to distort the base-ring or observing slit by using badly cut parts. It is not easy to preserve an accurate form when working on a complex shape, and regular checks with a tape measure and try-square are essential as each rib is installed – any twist or bend should be rigorously eliminated by reshaping or repositioning the part as required.

After about a week of evenings fitting the ribs, the framework was ready for skinning. This was done with 6 mm ($\frac{1}{4}$ in) exterior plywood and needed a total of about ten sheets to complete. The polygonal shape of the dome made this work quite easy, as master panel shapes were easily made and copied, and a small amount of planing would generally get the panels to the right profile in a few minutes. Three levels of tiling, each with eleven trapezoidal panels, were needed to completely skin the main body of the dome, and each panel was fixed in place with brass round-head screws and plenty of brown acrylic mastic. A liberal "sausage" of mastic was applied all around the rib edges where a panel was to be fixed, and then 30 mm ($1\frac{1}{4}$ in) brass screws were driven through the panel edges at intervals of about 300 mm (12 in), and tightened into the ribs below. The lowest circle of (rectangular) panels was fitted first, so that the upper ones could overlap in the required manner for rain proofing, and these were cut to a length that overlapped the support platform to prevent rainwater from entering the golfball track. The second and third echelons then followed, each with a 25 mm (1 in) overlap to the row below. The cutting and fitting of the entire skin occupied about another week of evenings during a (fortunately) dry spell in July 1994.

The Dome Doors

By this time, the dome was beginning to look rather smart and functional, but one major headache remained – how to fit the observing slit doors? I had originally intended to use some form of simple hinged arrangement, but the polygonal shape was rather too complex to allow for a simple "barn door". I eventually decided that the only practical arrangement would be a "Mount Palomar-style" two-leaf door (see Figure 9.6), which could roll back on metal tracks and small trolley wheels to either side of the slit. This, however, meant the construction of two very long polygonal arcs of woodwork and I was concerned that they would be too flexible to stay in shape and roll smoothly. There were two solutions to this:

Figure 9.6 The dome doors open showing the Tri-schiefspiegler and deep-sky camera.

1. The centre ribs, where the two doors met, would be broad and rigid. This would prevent the doors from opening beyond the last 18 mm ($\frac{3}{4}$ in) of the slit aperture, as the wide ribs would hit the slit walls, but this was a minor sacrifice for rigidity.
2. The cross-ribs of the doorway could be allowed to slide on Teflon-coated support blocks at various heights up the slit edges, so that the door panels would be supported along their lengths.

With these ideas in mind, I cut four polygonal arcs of 18 mm ($\frac{3}{4}$ in) ply, two narrow 50 mm (2 in) ones for the outer door edges, and two wide 150 mm (6 in) ones for the central ribs. Extra sections of ply had to be added to these ribs to make up the length required, and these were dovetailed on, as before. The doors needed to be shaped so that the rain would run away from the central joint when closed, and so the cross-ribs were cut with a 5° slope towards the outer ends. These were then used to join the inner and outer vertical ribs at intervals of about 500 mm (1 ft 6 in), the whole structure being screwed and glued together as strongly as possible. Because of the need for an accurate fit, both door frames were bolted together in the "closed" position and then hoisted into place over the observing slit. A 25 mm × 25 mm (1 in × 1 in) U-section aluminium channel was screwed onto the dome slit at each of the top and bottom ends to act as tracks for the door wheels, and pairs of 100 mm (4 in) diameter rubber-tyred plastic wheels were fitted to the equivalent points on the door frame. Once the wheels had been dropped into the tracks, the whole door assembly was checked for free running and adjusted as necessary.

The final operation was to skin the door halves in the same way as the dome itself, making sure that the mating edges were accurately parallel so that the gap was as small as possible when closed. I wanted the doors to have a neat appearance, and took a chance on the need for a sealing strip between the two halves, relying on the slope of the door skin to guide water away from the joint. This has proved to be quite satisfactory, even the heaviest rain being excluded very well by the combination of drainage angles. During bad weather, the doors can be bolted together, using 8 mm ($\frac{5}{16}$ in) bolts and wing nuts, but most of the time it is adequate to simply push them together, with even the strongest winds having no tendency to lift or separate them. I will probably motorise the opening and closing of the doors at some time in the future, but they are easily operated by hand and there is no urgency to upgrade the system.

Trouble with the Ballrace

When the dome was completed and operational, it quickly became clear that there was a serious problem with the golfballrace. The track contained about three hundred reclaimed balls collected from a local golf course and, although they would roll individually without trouble, once they began to bunch together there was a strong tendency for the balls to climb over each other and to force their way out of the track. I tried various ideas, such as breaking up the long groups of balls with sliding wooden blocks, but none was totally successful until a complete cage assembly was introduced. Had I looked closely at a modern ballrace, I would have realised that a metal cage assembly is used to hold the balls at regularly spaced intervals all around the track.

With the present design, no ball-bearing ever comes into direct contact with another. The natural roughness of golfballs makes it even more important that the balls are separated, and so I devised a simple caging technique, using 6 mm ($\frac{1}{4}$ in) hardboard. About ten 100 mm (4 in) wide arcs of the board were cut with a radius equal to that of the ball track, and were drilled at 100 mm (4 in) intervals with 45 mm ($1\frac{3}{4}$ in) diameter holes. A projecting tab at one end of each arc was fitted with a pin to engage in a hole in the tail end of the previous arc, so that the series of pivoting pieces became a complete circle when all had been assembled.

The dome was jacked up at intervals as each arc was slid into place and loaded with golfballs in the holes, until all of the free-rolling balls had been caged in the hardboard ring. Far fewer balls are needed in this new arrangement, and the tendency to ride up has been completely eliminated. The ballrace now works well and I can strongly recommend it to other dome constructors.

Finishing Touches

The main problem with any observatory is to combat the damaging effects of the weather. This is especially so when the building material is wood, as rot can easily set in and ruin many weeks of hard work. I have tried many different protective coatings over the years, and many can be rejected as quite unsuitable for a building which is exposed to bright sunshine and heavy rain and snow at regular intervals. Most of the damage is caused by ultraviolet light, which will destroy most polyurethane-based varnishes in a matter of months. Epoxy coatings are far more durable, but very expensive in the large amounts required, and so my favourite solution has proved to be the new water-based wood-protecting stains.

I have painted the dome with four coats of gloss Ronseal Woodstain, which has resisted the weather remarkably well, to date. The finish is a mid-brown wood-grain which merges with the surroundings very well, and does not seem to cause any significant thermal problems for the telescopes. It is important to varnish the inside of the dome as well, as the bare wood is likely to develop mildew spots in damp weather, and can quickly deteriorate. One coat is generally enough to prevent problems, but can be difficult to apply to the upper surfaces. This is best overcome by the use of paint pads, rather than brushes, as they can be loaded with a lot of varnish without dripping and applied with a long broom handle to reach the top of the dome.

Would I Do Anything Differently Today?

To be frank, I cannot think of anything that needs to be drastically altered, although the observing slit would benefit from an increase in width. This is currently about 950 mm (3 ft 1 in) wide, and is okay for objects close to the meridian; however, the double-barrelled camera-and-telescope combination tends to bridge the slit width when observing towards the eastern or western horizon, necessitating frequent dome rotations. In all other respects the observatory serves its purpose very well and has enabled me to capture many very satisfactory CCD images of the planets and deep-sky objects.

Chapter 10

The University of Hertfordshire Observatory in Bayfordbury, England

C.R. Kitchin

Introduction

The University of Hertfordshire Observatory (UHO) was founded in 1970 to provide observing opportunities for astronomy students at what was then the Hatfield Polytechnic. The polytechnics were a new development in higher education in the United Kingdom, being intended to provide generally industrially biased degree courses, frequently of a "sandwich" nature. The inclusion of a pure science like astronomy was therefore unusual. In fact, among the fifty or so polytechnics only three included astronomy in their portfolios from the start.

The start of astronomy at Hertfordshire was largely due to the efforts of one man, the first Director of the UHO; J.C.D. (Lou) Marsh. He was already working as a lecturer in electrical engineering at the college when he put on a trial course of lectures on general astronomy. This was received with great enthusiasm by the students and so in 1967 he proposed to the academic board that astronomy be offered as a regular subject, and that an observatory be built to provide support for the courses. With the help of several members of the top management of the college who had personal

Figure 10.1 General photograph of the observatory site showing the dome of the Marsh telescope on the right, that of the spectroscopic telescope on the left and those of the 0.35 m (14 in) Schmidt–Cassegrain and Brinton and Vince telescopes in the distance. In the foreground are Dr Chris Kitchin (Observatory Director, *left*) and Iain Nicolson (Astronomy Lecturer, *right*).

interests in astronomy (the Director, Sir Norman Lindop, went on to take an MSc in Astronomy and Astronautics at the UHO after he had retired), the proposal succeeded. The first formal courses in astronomy were thus taught in 1969.

Funds were also provided to acquire a telescope and to build a dome for it.

The college itself is poorly sited for an astronomical observatory, being within the town of Hatfield, and close to several major roads, a railway and an airfield, and with several other sizeable towns nearby. Luckily, it possessed an annex in the grounds of a country mansion some ten miles away, which had far less light-polluted skies. The observatory was therefore sited at this annex, Bayfordbury, where it continues to this day.

The first telescope was a 0.4 m (16 in) Cassegrain-Newtonian on a modified English mounting, produced by a local firm called Astronomical Equipment. It was housed in a 5 m (16.5 ft) Ash dome, and was opened by Dr Alan Hunter, the Deputy Head of the Royal Greenwich Observatory, in 1970.

Since 1970 the observatory has developed considerably, with numerous telescopes, four full-time members of the teaching staff, research staff, technician support, and so forth, and now provides observing opportunities for students on a wide range of astronomy courses. The Hatfield Polytechnic became the University of Hertfordshire in 1991, and so the HPO then became the UHO. 1995 saw the twenty-fifth anniversary of the opening of the observatory, and the opportunity was taken to name that first telescope, now upgraded to a 0.5 m (20 in) Cassegrain, after the first Director, and it is thus known as the Marsh telescope.

The Domes

The Ash Dome

The original dome, now housing the Marsh telescope (see Figure 10.2), was brick built with a 5 m (16 ft 6 in) dome. That has continued in use with very few problems since 1970. A few minor modifications have been made, in particular to improve the automatic end-stops on the shutter drive. Those originally supplied were too small, and the shutter often overran the top stop, necessitating someone climbing onto the north pier of the telescope and leaning precariously

Figure 10.2 The Marsh telescope.

out to wind it back by hand, at great risk to life and limb!

The azimuth drive motor for the dome has had to be rewound on a couple of occasions, and in 1995 a rack-and-pinion drive was substituted for the original friction drive which had worn and begun to slip.

As further telescopes have been acquired, new means of housing them have been needed. Unfortunately, since that first dome, funds have never been adequate for another brick building.

Numerous alternatives have therefore been tried over the years, several of them being constructed by the observatory staff. Those alternatives (and their fates, many of which readers may want to class in the "dire warnings" category) have included:

A Wooden Run-off Roof

A 3.6 m (12 ft) square wooden building with a run-off roof (see Figure 10.3). This was based on 150 mm (6 in) posts and beams and 12 mm ($\frac{1}{2}$ in) exterior grade plywood. The roof weighed some 250kg (0.25 tons),

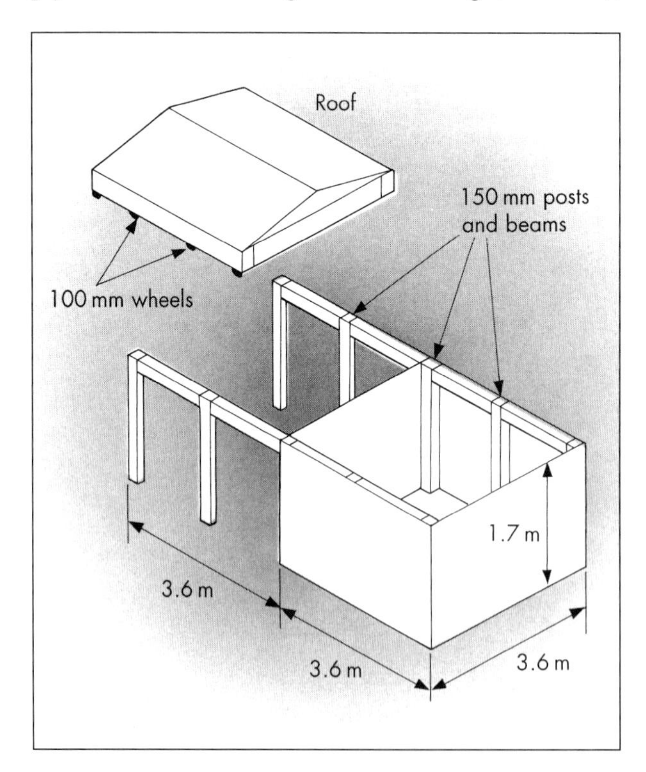

Figure 10.3 Split diagram showing the construction of the run-off roof observatory.

and ran on 100 mm (4 in) wheels. Despite its weight, and despite being held down by steel hooks, the roof was blown off in a gale one day and totally destroyed.

A Glass Fibre Shelter

A 2 m × 3 m (6 ft 6 in × 10 ft) glass fibre workman's shelter was mounted on wheels which in turn ran on rails. The whole shelter ran back to allow the telescope to be used. This shelter also blew away (twice!) in gales, but survived intact. It was however inconvenient, and left the telescope unprotected from the wind when in use. It was therefore eventually abandoned. The 250 mm (10 in) Newtonian telescope which it housed was remounted on the side of a 300 mm (12 in) Newtonian to provide a twin telescope.

Glass Fibre Domes

The observatory has used two fibreglass domes, one was a 3 m (10 ft) dome mounted on a separate 3.5 m (4 ft) high wall cylinder, the other, also 3 m (10 ft) in diameter, had the dome integral with the walls and the whole structure rotated on rails. The former dome was lifted off its wall cylinder by the wind on three separate occasions, and eventually was irretrievably damaged, the latter dome disintegrated completely in a particularly severe gale.

Wooden Domes

The observatory has tried three domes made of plywood/hardboard, all 3 m (10 ft) in diameter. Two were on 1.5 m (5 ft) wall cylinders and one on a 3 m (10 ft) high wall cylinder. These have performed better than the glass fibre domes, but nevertheless the first two were damaged beyond repair in the same storm that caused the second glass fibre dome to disintegrate.

A problem with all three has been that the shutters were just sheets of plywood running in narrow slots. These either jammed or came out of the slots in use. The one dome of this type that is still used has there-

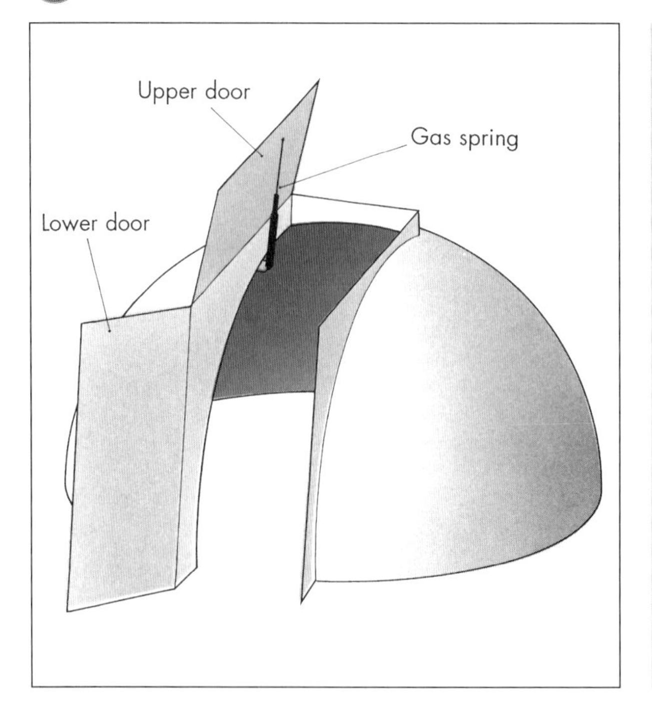

Figure 10.4 Design for the replacement doors for the original shutters in a plywood dome. The upper door is operated by a gas spring, which pushes it up and is then pulled down again by a rope.

fore had the original shutter replaced by aluminium doors (Figure 10.4). That dome is now about 15 years old, and in addition to the replacement shutters has had to have its panels replaced once.

It may well appear to the reader, from this catalogue of disaster, that the observatory is either sited in an extremely windy site or that the staff are remarkably careless in caring for the domes (or both).

Neither is in fact the case, but our experience shows the need for very rugged construction indeed if a dome on an exposed site is to last any length of time. Gales of force 10 on the Beaufort scale are rare inland, but do occur at most sites once or more over a five-year interval. In such a gale the average wind speed is about 100 km/h (60 mph), but gusts can reach 150 km/h (90 mph). The wind force at 150 km/h is nearly 120 kg/m^2 – over a tonne on a 3 m (10 ft) dome on a 1.7 m (5 ft) wall cylinder!

Any dome or other housing for a telescope must thus be designed with these sorts of wind forces in mind.

Nowadays, therefore, the telescopes at the UHO are mostly housed in galvanised steel domes produced by Ash Domes, and mounted on substantial wall cylinders. The wall cylinders are based on 150 mm ×

50 mm (6 in × 2 in) uprights, surrounded by 12 mm ($\frac{1}{2}$ in) exterior-grade plywood and then corrugated aluminium sheeting (see Figure 10.1). They have a rigid steel jacket at the top which incorporates the track for the dome, and are securely bolted into a thick concrete base.

Instrumentation

As I mentioned in the introduction, the observatory started with a 0.4 m (16 in) Cassegrain-Newtonian telescope. In 1985 that instrument was replaced with a 0.5 m (20 in) Cassegrain on the original mounting, and this is now called the Marsh telescope. A 0.15 m (6 in) refractor acts as the guide telescope, and a 0.2 m (8 in) Schmidt camera is piggybacked onto the same mounting.

A new secondary was obtained for the original 0.4 m (16 in) mirror to give an f/9 final focal ratio in order to match it to a Optomechanics spectroscope. A new telescope tube and a fork mounting were obtained for this mirror and the whole installed in an Ash dome to provide a dedicated spectroscopic system (see Figure 10.5).

Also permanently mounted in domes are:

Figure 10.5 The 0.4 m spectroscopic telescope with the spectroscope installed.

(i) A 0.4 m (16 in) Meade computer-controlled Schmidt-Cassegrain telescope
(ii) A 0.36 m (14 in) Celestron Schmidt-Cassegrain telescope which has a 0.2 m (8 in) Celestron as a guide telescope, and which has been provided with a new and more robust fork mounting (see Figure 10.6).
(iii) A 0.3 m (12 in) Newtonian telescope built at the turn of the century and donated to the observatory in 1976 by Henry Brinton
(iv) A 150 mm (6 in) refractor with a Mertz objective donated by Phillip Vince.

In addition to these permanently installed instruments, four 0.2 m (8 in) and four 125 mm (5 in) Schmidt-Cassegrain telescopes are used as portable instruments. Students learn how to use telescopes on these smaller instruments before moving on to the larger ones. Another six 125 mm (5 in) Schmidt-Cassegrain telescopes are housed in an astronomical laboratory to provide preliminary training and for cloudy-night work on simulations.

A 0.2 m (8 in) coelostat feeding a fixed 0.2 m (8 in) Schmidt-Cassegrain telescope for solar work, and 5 m and 3 m ($16\frac{1}{2}$ ft and 10 ft) radio dishes complete the primary instrumentation of the observatory.

Figure 10.6 The 0.36 m Schmidt–Cassegrain Telescope.

When the observatory opened, photography was the only imaging technique in use. Now although a small amount of work is still done with photography, mostly for the Schmidt camera, it has been superseded for general imaging by CCD detectors. The observatory currently has eight of these, two from EEV, five from SBIG and one from Sanyo. Six optical photometers and four visual micrometers are also used, and an imaging polarimeter is nearing completion. For solar work, most of the telescopes are provided with Solar Skreens, and there is an H-alpha filter.

For laboratory work, there are two microdensitometers, and a 2-D plate-measuring machine, together with numerous computers for processing images, for running data reduction programmes and for simulations. There is a also a large archive of UK Schmidt plates available for use.

Practical Work by Students

Practical work at the observatory is expected of any student studying astronomy at the university. On clear nights an appropriate observing programme will be assigned to individual students or to small groups of students using the above instrumentation. On cloudy nights, students work in the laboratory, perhaps processing previously obtained data, or using 125 mm (5 in) Schmidt-Cassegrain telescopes to make observations of simulated objects (the observatory has an in-house-produced computer program providing a simulated star field, with a dozen variable stars of different types, a galaxy, globular cluster, etc.). They may also use material obtained by staff on major telescopes such as the Isaac Newton telescope, International Ultraviolet Explorer (IUE) spacecraft, etc., or use some of many UK Schmidt plates stored at the observatory.

Research

Although the observatory is a teaching facility, the astronomy group has an active research role. The main research areas are into active galactic nuclei, and young stellar objects and are strongly supported by the Particle Physics and Astronomy Research Council (PPARC). Observations are obtained on instruments such as the William Herschel telescope, the UK Infrared Telescope (UKIRT), the Anglo-Australian telescope, etc., and processed on the STARLINK node at the university. Many of the observations involve polarimetry utilising the polarimeters developed at the university, but now available as common-user instruments. There are currently seven members of staff and three research students actively working in these areas. Frequently undergraduate students are able to join in this research as part of their project work using STARLINK software on University computers to process some of the research data.

Public Activities

The observatory plays a major role in bringing science to the public both in the locality and to a wider audience. The extramural courses have already been mentioned. Two general open evenings per year attract up to 500 people on each occasion, and when not in use for the university's students, the observatory is open by prior arrangement to visits from local schools, astronomy societies and other such groups.

The staff of the observatory publish regularly in popular astronomy journals such as *Astronomy Now*, and give invited lectures in schools, astronomy societies, etc. A number of books have been published by the staff at levels ranging from the popular to research monographs. Staff also appear regularly on local and national radio, and on television in such programmes as *The Sky at Night* and *Heavenly Bodies*. The observatory is used to provide observational experience to students on the Open University's course "Astronomy and Planetary Science". Other areas of involvement include "Astrofest" and the National Science and Technology Week.

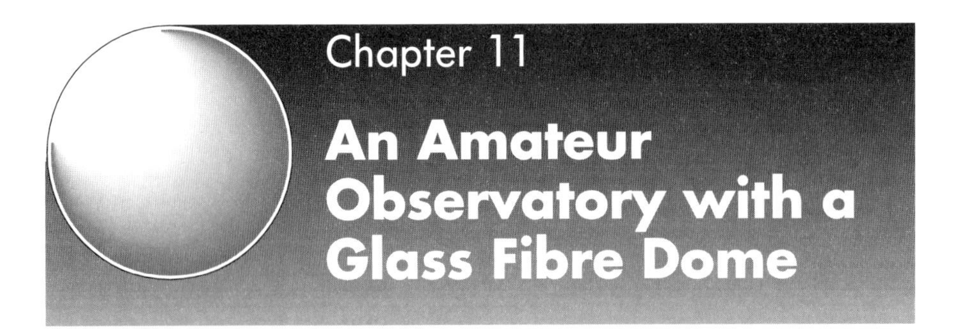

Chapter 11

An Amateur Observatory with a Glass Fibre Dome

Ron Johnson

Introduction

When I first started using a telescope I spent many hours observing in the open air. During the winter months it only took about half an hour before the cold started to penetrate the layers of clothing I had

Figure 11.1 The complete observatory. The telescope tube is also visible.

on. On one occasion my eyebrow froze to the eyepiece!

Condensation on the telescope and optics was also a problem from time to time. The telescope was permanently mounted, which meant that it had to be uncovered and re-covered before and after each observing session. Accessories had to be taken from the house to the telescope and returned again afterwards. All these pre- and post-observing activities took quite a time, time that could be better used observing.

Design Philosophy

Building an observatory was the obvious answer, as it would provide protection while observing, as well as space for keeping all the accessories. The design needed to be simple and easy to construct. I am not a craftsman but can at least manage to use a few simple tools: a hammer, saw, drill and adjustable spanner were the main tools I needed.

The size of the observatory was determined by the size of the telescope it had to accommodate. I had already constructed a 297 mm ($11\frac{3}{4}$ in) reflector, which was the instrument to be housed in the observatory. Allowing for a small storage area and space for two or three observers I decided that a 3 m (9 ft 10 in) observatory would be needed.

After considering several different outline profiles for my observatory I decided to make it square (plan view). Circular observatories are more pleasing aesthetically but square ones are easier to construct and furnish, and the corners make good storage areas.

A major design consideration was that the observatory had to be demountable so that if I moved house it could be dismantled and reassembled in another location. The walls were therefore framed up in timber panels and bolted together. The flat corner roof areas were also in timber. Right from the beginning, I wanted to incorporate a fibreglass dome as it would look good, be light in weight, extremely strong, easy to operate and need very little maintenance. I realised that the dome would have to be made in panels and bolted together. This, then, gave me the basis for the development of the design.

Construction

The base of the observatory consists of a reinforced concrete edge-beam 200 mm × 300 mm (8 in × 12 in) in section, and 3 m (9 ft 10 in) square (on plan) which supports the timber walls. Holding-down bolts 150 mm (6 in) long were cast into the top of the edge-beam to take the wall panels.

The eight panels were prefabricated using 100 mm × 50 mm (4 in × 2 in) timber framing with members nailed together. All the timber was creosoted prior to assembly. It is particularly important to remember that all cut ends are properly treated. Roofing felt was then nailed to the outside face of the framing which was then clad with creosoted 150 mm × 19 mm (6 in × $\frac{3}{4}$ in) shiplap boarding. The felt prevented any draughts penetrating through the timber cladding.

A damp-proof course was laid on top of the concrete edge-beam before erecting the wall panels, which were bolted to the beam and to each other. A timber plate 100 mm × 50 mm (4 in × 2 in) was nailed to the top of the wall panels along each side of the observatory which made a very rigid structure.

Figure 11.2 The observatory walls completed, together with the flat roof areas and internal timber upstand.

Because the dome is circular and the walls are square, there is a small area of flat roofing required at each corner of the observatory. This was formed by fixing 100 mm × 50 mm (4 in × 2 in) diagonal bracing across each corner and then nailing timber boarding over the top of the walls and the bracing and then trimming it to fit. The corner roofing now formed an octagonal shape on plan around the inside of which 150 mm × 19 mm (6 in × $\frac{3}{4}$ in) timber was fixed to form an upstand (see Figure 11.2). Roofing felt was then laid on the boarding and up the outer face of the timber upstand.

The design of the dome was such that only two different panel sizes were required, one for the side panels and one for the fixed and removable shutter panels. Traditionally, domes have panels tapering from bottom to top. This involves a lot of awkward cutting to form the observing aperture, and few of the panels are the same size. Making the panels taper from top to bottom eliminates all the cutting, and forms the observing aperture automatically.

It is essential to have an observing aperture that is large enough to enable you to observe without having to move the dome every few minutes. I find an aperture of about 1 m (3 ft) is adequate for a 3 m (9 ft 10 in) dome.

Once I had established the dimensions of the dome, I divided it up into manageable panel sizes that could be easily constructed in the garage space available. This resulted in five panels on each side of the dome, and five shutter panels that went over the dome.

The colour of the dome is important, because it should reflect heat from the sun but should blend in with the background environment; requirements that may conflict. A useful way to establish which colours are suitable is to cut some table tennis balls in half and paint them a selection of different colours. Then place them on a board against the background where the observatory is to be sited and view them from a distance. Dark colours are best avoided as they generally absorb more heat than lighter colours. White may be considered an ideal colour but it is very bright when the sun is shining on it. It also requires regular cleaning.

In the end I compromised and decided on a light grey. This blends in with the sky background and only needs to be washed every few years.

Making the Glass Fibre Dome

To construct the dome panels two glass fibre master moulds were needed, one for the side panels and one for the shutter panels. To make the master moulds, I constructed two formers out of timber and plaster. It was not hard to cut and shape the timbers to form the curve in one plane, but plaster was needed to form the curve in the other plane. Timber framing covered with hardboard was constructed to predetermined dimensions and then plaster was added and shaped by using a curved running-mould. Once set, the plaster was sealed and the whole former painted with a release agent to prevent the glass fibre sticking to the timber and plaster.

The process of glass fibre construction involves laying down alternate layers of resin and glass fibre matting until the required thickness is achieved. A hardener or accelerator is added to the resin before you use it, to make it harden. The more you add the quicker it hardens! Temperature will also affect the time it takes for the resin to harden; the warmer it is the quicker it will harden. It is essential to work in a well ventilated area and to wear protective clothing.

The master moulds needed to be strong as they were going to be used several times, so they were made up using three layers of 600 g/m^2 (2 oz/ft^2) glass fibre matting. It is easier and quicker if laying the glass fibre is carried out by two people. Each master

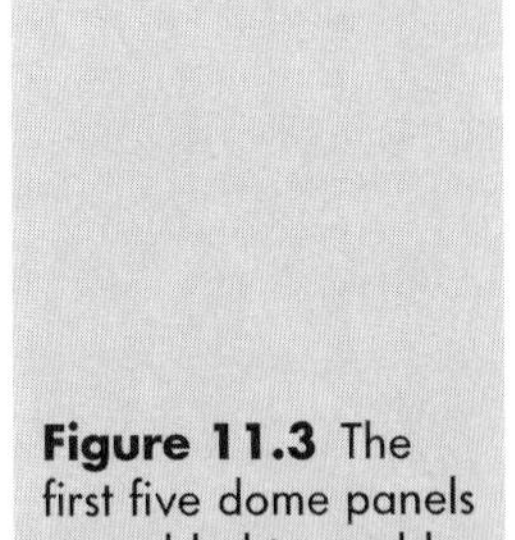

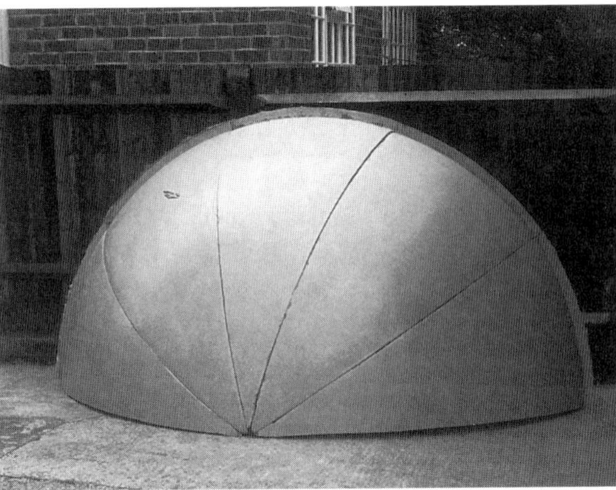

Figure 11.3 The first five dome panels assembled to enable overall dimensions to be checked prior to erecting the panels on the observatory.

mould took about two hours to make (with the assistance of a colleague), and was left overnight to cure.

It is essential that each master mould has at least one removable side, which can be made of 12 mm ($\frac{1}{2}$ in) plywood or chipboard. This allows you to release the panels from the mould. When the master moulds had hardened and were taken off the formers, they were trimmed up and the inner surfaces buffed up to a high gloss using slipwax, which also assists in releasing the panels from the mould.

The moulds were now ready to be used to produce the dome panels.

Each dome panel was cast in the same way. First we gave the mould a coat of release agent, which was allowed to dry. The first resin coat (the gel coat) was mixed with a pigment to give the panel its colour and then applied to the mould. The exposed surface of the gel coat remains tacky, so as to form a good key with following coats. A layer of 600 g/m^2 (2 oz/ft^2) glass fibre matting was then laid on top of the resin and rolled with a laminating roller to make sure that no air was trapped between the resin and the matting. The resin should totally penetrate the matting.

Once you have done this, you add another resin coating, followed by a further layer of matting. I used two layers of glass fibre matting in all the dome panels. After each panel was released from the mould I trimmed its edges with a hacksaw.

As production of the dome panels progresses, it is a good idea to bolt the panels together on the ground as a check that the overall dimensions are correct, prior to final assembly (see Figure 11.3).

Assembling the Observatory

Final assembly is easy (see Figure 11.4), but remember to place a strip of flexible sealant between each panel before tightening up the bolts. As the bolts are tightened up the sealant is squeezed between the panels to make a watertight joint.

Prior to positioning the dome on the observatory, the base-ring had to be set up. The base-ring of the dome consists of a 50 mm × 50 mm (2 in × 2 in) mild steel angle, radiused to suit the dome. It runs on six

Figure 11.4 (*opposite*) Observatory assembly: dome and shutter panels.

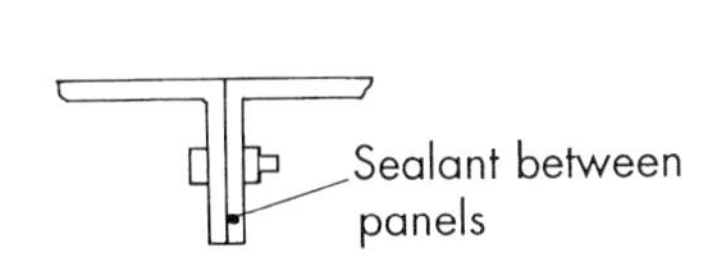

Bolted connection of adjacent dome panels

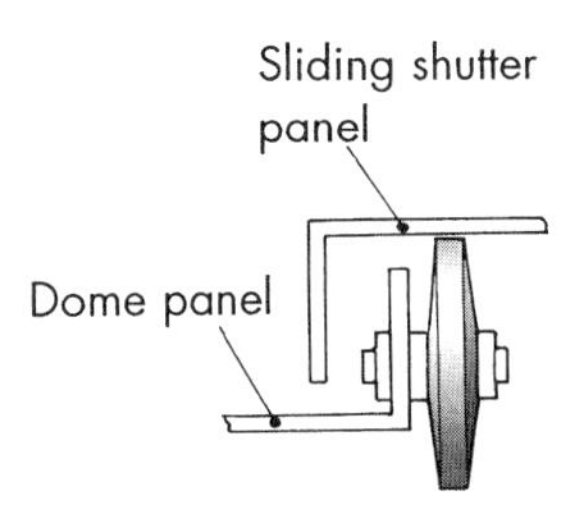

Plastic wheel fixed to dome panel to assist in sliding the shutter panel over

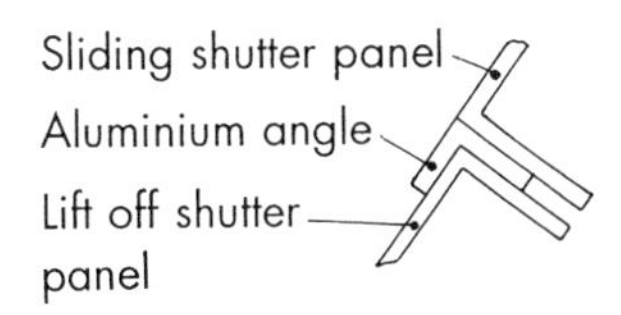

Joint between sliding and lift off shutter panels

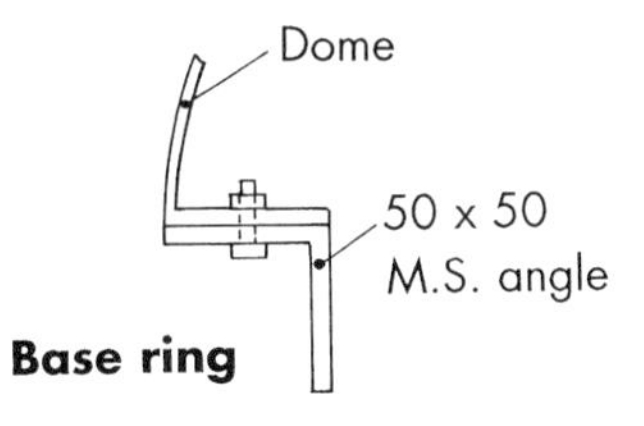

Base ring

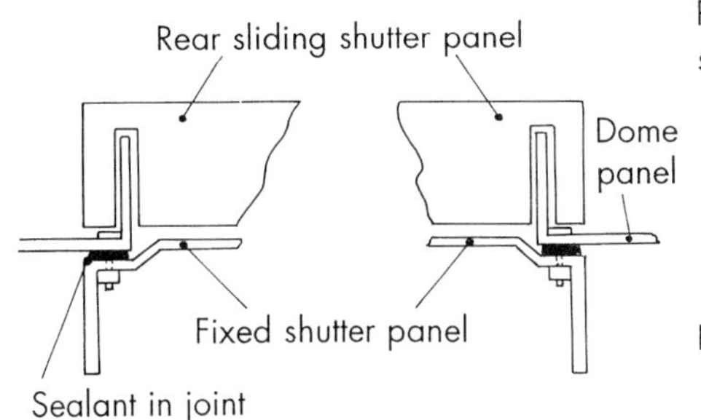

Rear view of sliding shutter panel

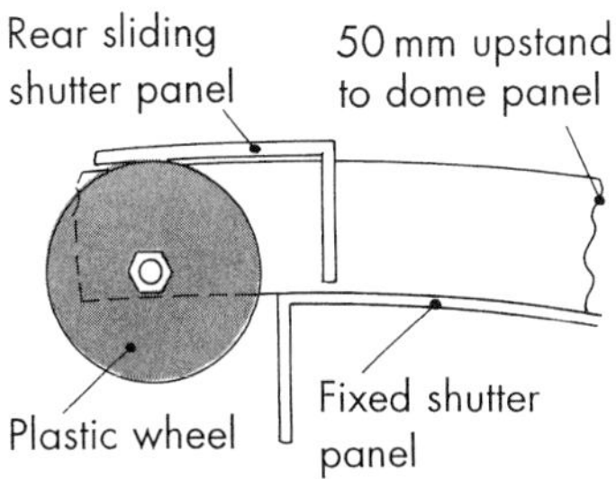

Section through fixed and sliding shutter panels

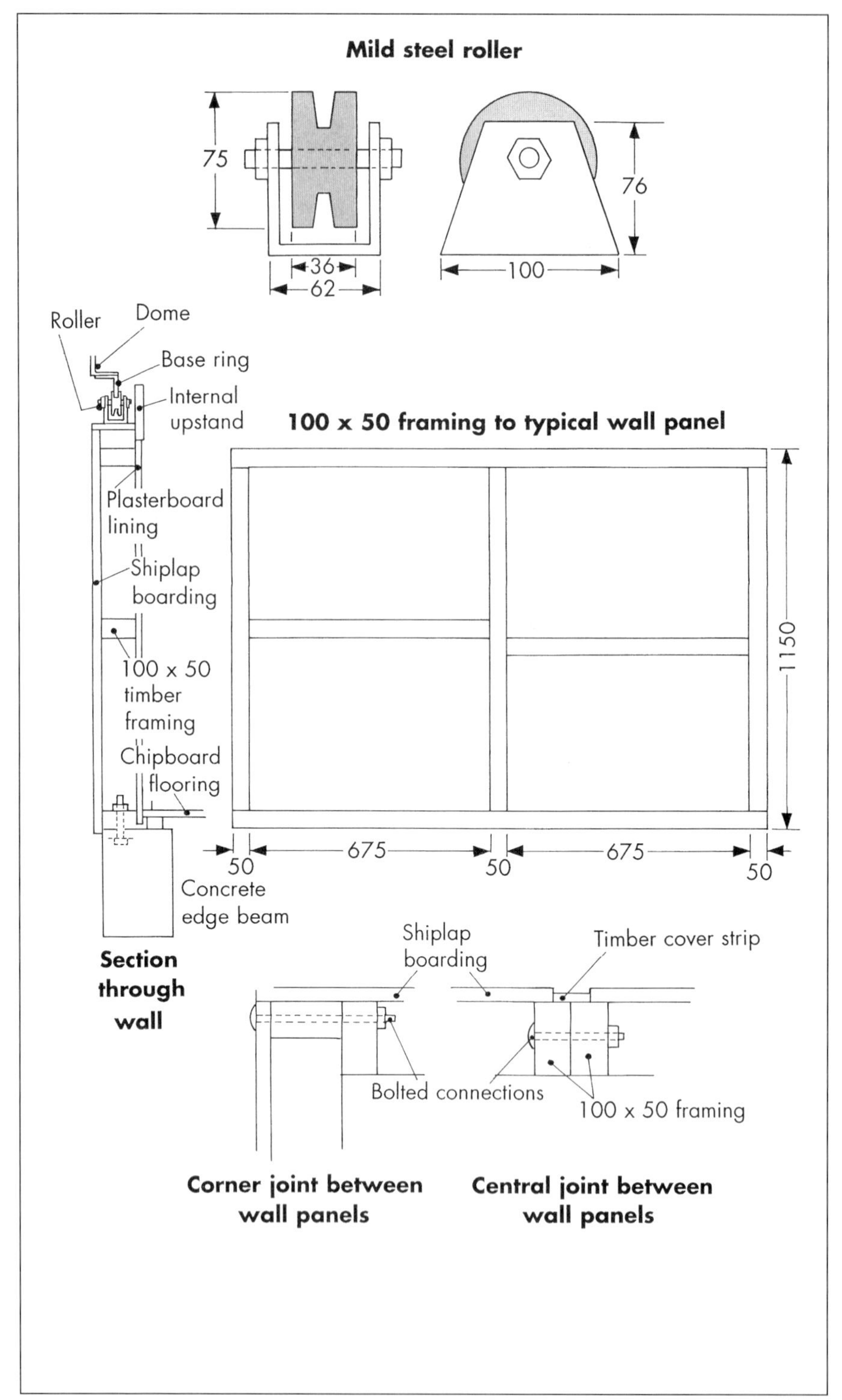
Mild steel roller
75
36
62
76
100
Roller
Dome
Base ring
Internal upstand
100 x 50 framing to typical wall panel
Plasterboard lining
Shiplap boarding
100 x 50 timber framing
Chipboard flooring
1150
50
675
50
675
50
Concrete edge beam
Section through wall
Shiplap boarding
Timber cover strip
Bolted connections
100 x 50 framing
Corner joint between wall panels
Central joint between wall panels

recessed rollers fixed to the flat roof areas of the observatory (see Figures 11.5 and 11.6).

The base-ring was made in two halves, bolted together. The idea was to make it easy to dismantle and transport if required.

Once the base-ring was completed, I had to lift the first five dome panels into place. It is at this point that you need some friends to help with the lifting, although the panels are more bulky than heavy.

The second five panels were then positioned, and then the fixed shutter panels were bolted in position. All the panels were then bolted to the base-ring. Finally, the remaining shutter panels were placed in position.

If the rollers are kept well greased the dome can easily be pushed round using one hand. The lowest front shutter panel lifts off if required, and the two panels above it slide back over the dome to give full access to the zenith.

Some form of wind restraint is needed to prevent the dome from being blown off in high winds. I tend to rope mine down at the moment, but a more permanent solution would be better.

The Telescope Mounting

Once the dome was in position the observatory was watertight, which allowed work inside to progress. I could now construct the telescope mounting. I chose to use an "English", or yoke, mounting (see Figure 11.7); the main reason was that it was easy to construct using only a few simple tools.

The yoke consisted of 150 mm × 75 mm (6 in × 3 in) timber on the long sides and 225 mm × 75 mm (9 in × 3 in) timbers on the short sides. Mild steel plates were bolted to each end of the yoke on to which were welded 37 mm ($1\frac{1}{2}$ in) diameter steel shafts (the polar axis). The shafts run in self-aligning bearings, supported on the north and south piers of the mounting.

Figure 11.5 (*opposite*) Observatory assembly: wall panels.

Figure 11.6 The rollers upon which the base-ring runs.

Figure 11.7 The 297 mm (11.7 in) and the English mounting, viewed through the open dome. Note the base-ring of the dome at the bottom of the photograph.

The north pier consists of a concrete base approximately 1.5 m (5 ft) deep into which I cast a 150 mm (6 in) diameter cast iron rainwater pipe. I filled the pipe with concrete. This made a very solid structure to support the north end of the mounting. The south pier was also constructed in concrete, in the form of a base and shaped low pier (see Figure 11.8).

It is important that the pier bases are isolated from the floor of the observatory to prevent any vibration being transmitted to the telescope.

I made the steel plates and brackets that support the bearings with as much adjustment as possible designed into them. This took the form of elongated slots in the steel plates, and brackets which allowed both horizontal and vertical adjustment.

The declination axis consists of 25 mm (1 in) diameter steel shafts which are welded to pipe flanges fixed to the sides of the telescope. The shafts are supported by flange bearings fixed to the inside of the yoke mounting.

The telescope tube is built out of octagonal and circular 19 mm ($\frac{3}{4}$ in) and 12 mm ($\frac{1}{2}$ in) plywood with the centre cut out to a diameter a little larger than the mirror. Two thicknesses of 19 mm ($\frac{3}{4}$ in) ply were used at the bottom of the tube to support the mirror cell. The outer edge of the ply takes 50 mm × 25 mm (2 in × 1 in) timber members that run the full length of the tube. Extra stiffening was added to the sides of the tube to take the declination axis shafts. The outside of the tube is covered with 3 mm ($\frac{1}{8}$ in) ply which is

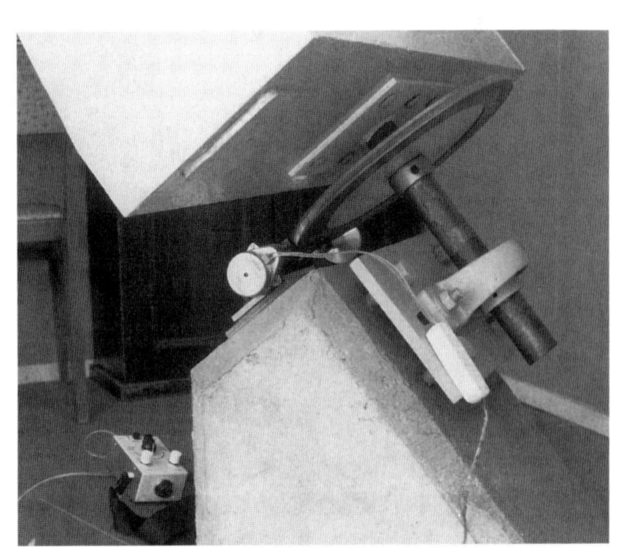

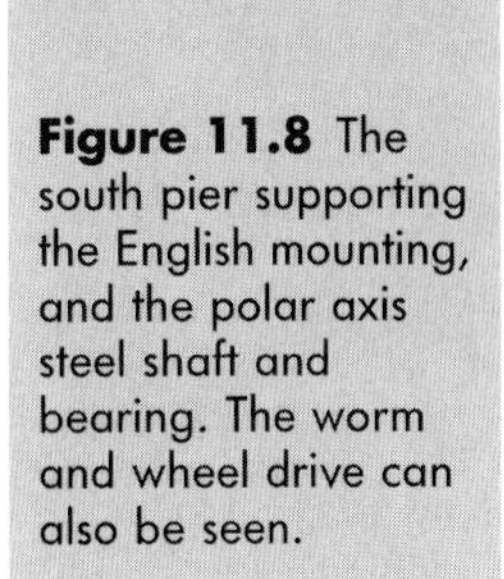

Figure 11.8 The south pier supporting the English mounting, and the polar axis steel shaft and bearing. The worm and wheel drive can also be seen.

pinned to the tube framework. The section of the tube adjacent to the declination axis is clad in 12 mm ($\frac{1}{2}$ in) ply for added rigidity.

After assembly, the mounting and tube were painted matt black on the inside of the tube to minimise any reflections.

The optics were then installed into the telescope and the focusing mount and finder added. The telescope then had to be balanced: this called for small amounts of lead, fixed in strategic positions.

The last part of the job was to fit a worm and wheel drive to turn the polar axis. An accurate drive is desirable if you are sketching the moon or planets, and essential if you are going to use the telescope for astrophotography.

Running an electrical supply to the observatory and any wiring should be carried out by a qualified electrician.

Post Mortem

The telescope and observatory have been in use for many years and everything has worked well. The observatory has needed little maintenance during this time.

Since I built my observatory, the cost of resins and glass fibre materials has increased considerably. This might influence my choice of materials for the dome if starting from scratch today, but, cost apart, the glass fibre dome has been entirely satisfactory.

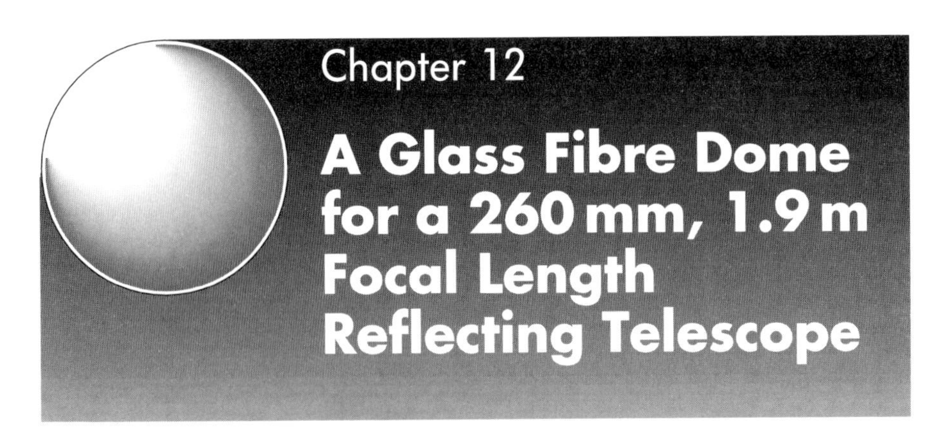

Chapter 12
A Glass Fibre Dome for a 260 mm, 1.9 m Focal Length Reflecting Telescope

B.G.W. Manning

For many years, my telescope, which is a Newtonian on an English-type equatorial mounting, had been housed in an asbestos sheet construction with a roof which opened by means of three hinged sections. *(Asbestos is now a banned material, and anyone replacing an old structure containing asbestos must get professional help in dismantling it and disposing of it.)*

Designing the Dome

When I moved house in 1966 I decided that a rotating dome would be nice. I considered constructing it in various materials, including glass-reinforced plastic (GRP), but I was not too happy about making the large moulds which would be necessary, and the trimming and sanding of the moulded sections. The possibility then occurred to me of making formers of expanded polystyrene sheet and covering both sides with a layer of GRP, which I discovered is a well-known method of construction. However polyester resin dissolves polystyrene, which means that the expanded polystyrene sheet must be sealed before the resin is applied. I did this by pasting on newspaper with wallpaper paste. Painting the polystyrene

Figure 12.1 Brian Manning's glass-fibre-domed observatory. Photograph © British Industrial Plastics Limited.

with (water-based) emulsion paint is supposed to work, but I found when I tried it that it leaves pinholes which allow the resin to leak through and dissolve parts of the polystyrene, which is fatal.

The polystyrene would of course be incorporated in the finished dome which should be quite light and yet rigid, and a further advantage would be that it would provide some insulation against the Sun's heat during the day.

Drawings and calculations were made (see Figure 12.2) and the details are as follows:

Inside diameter 3.2 m ($10\frac{1}{2}$ ft), centre of curvature 150 mm (6 in) above the base to allow for the telescope to aim nearly horizontal if ever required to do so. Slot width 0.86 m (34 in), extending 150 mm (6 in) beyond the apex, and covered by an up-and-over sliding shutter which terminates 300 mm (12 in) from the base. There is a removable panel to cover the gap (see Figure 12.3).

There are four main dome sections, three of which are fastened to the aluminium track on which the dome revolves. Two sections are approximately half-hemispheres, each consisting of an arch and six pa-

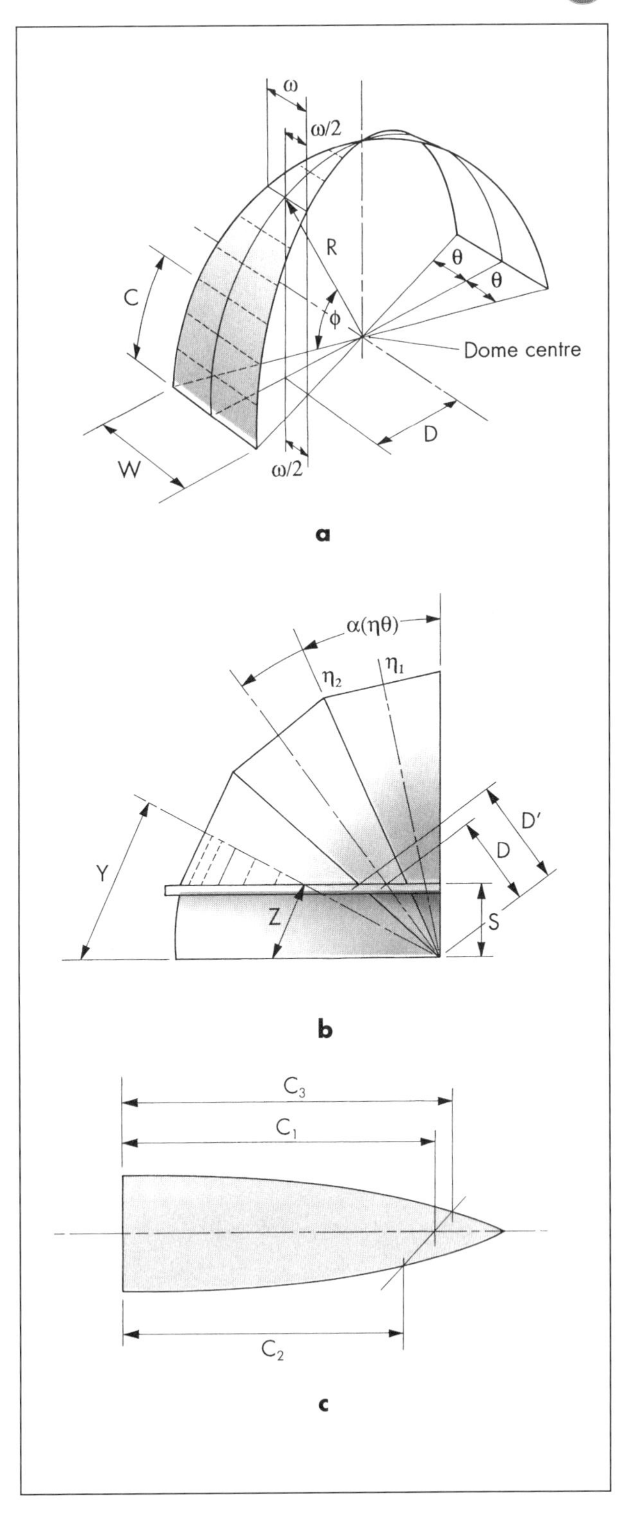

Figure 12.2 Each panel can be considered to be part of a cylinder whose axis passes through the centre of the dome, (part **a**), and the edges of the panels are defined by oblique sections of the cylinder. The outside width of the slot is 0.9 m (36 in) and the 12 panels each subtend an angle 2θ of 24.5 degrees. Maximum width of panel W is 2Rtan(θ).

I calculated the ordinates w at 10deg. intervals (part **a**). $C = \phi/360 \times 2\pi R$ ordinate width is $w = W\cos(\phi)$. This gives the basic shape of the panels; they are however intersected by the slot (part **b**). Two are only slightly truncated and the lengths C1, C2, C3 were calculated (part **c**), as follows. For C1: $D = S/\text{Cos}(a)$, (part **b**), $\phi = \text{ArcCos}(D/R)$, $C1 = \phi/360 \times 2\pi R$.

For C2 and C3 D′ is required. $D' = S/\text{Cos}(a)$. Where panels abut, $C2 = C3$.

Panel adjecent to slot, ordinates at 10 deg intervals as before. Z (part **b**) is first found. $Z = S/\text{Cos}(90-a)$, then $Y = R\,\text{Cos}(\phi)\text{Tan}(90-a)$, ordinate is Y–Z.

Allow for $\frac{1}{2}$ to 1% shrinkage on pasting and any extension below equator.

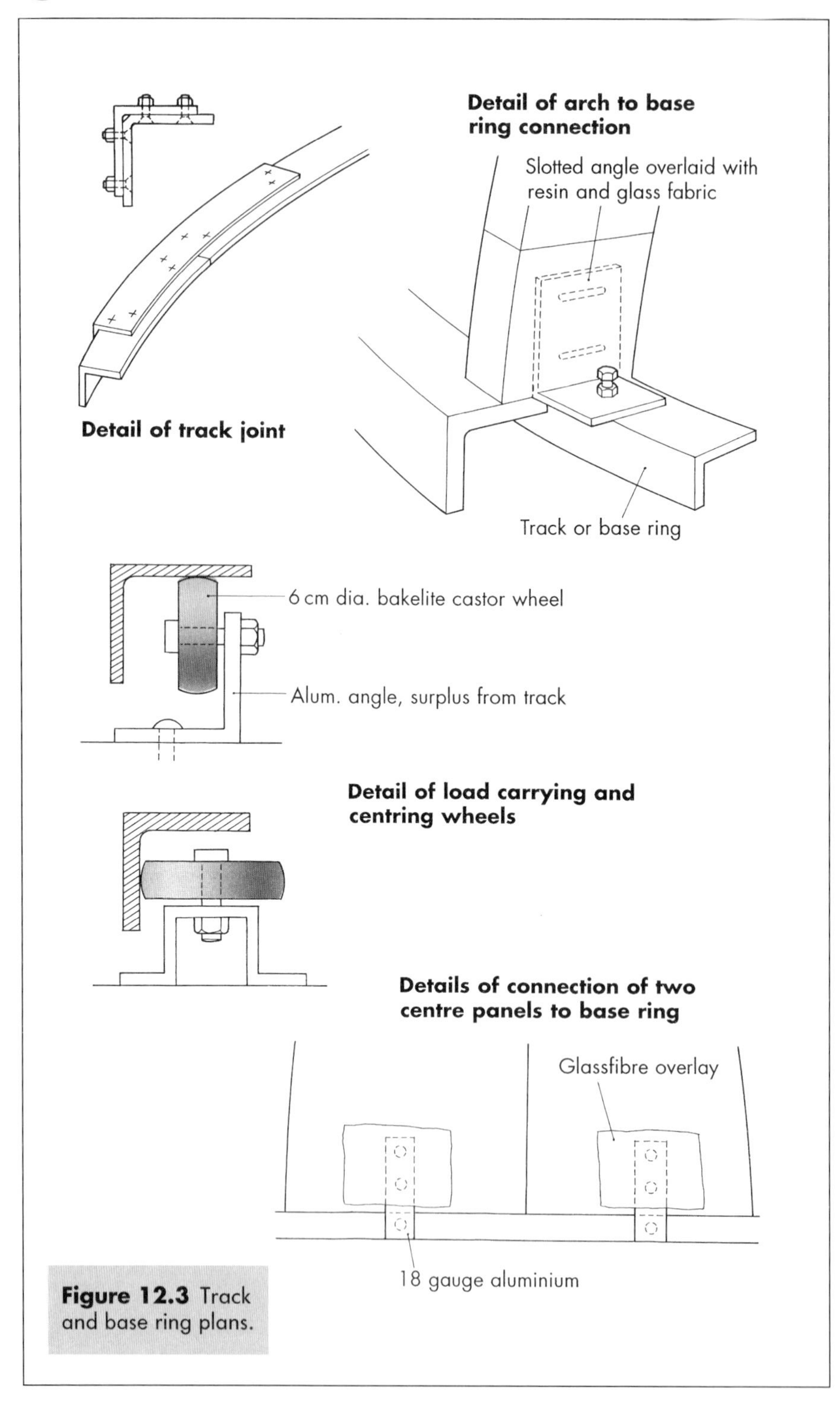

Figure 12.3 Track and base ring plans.

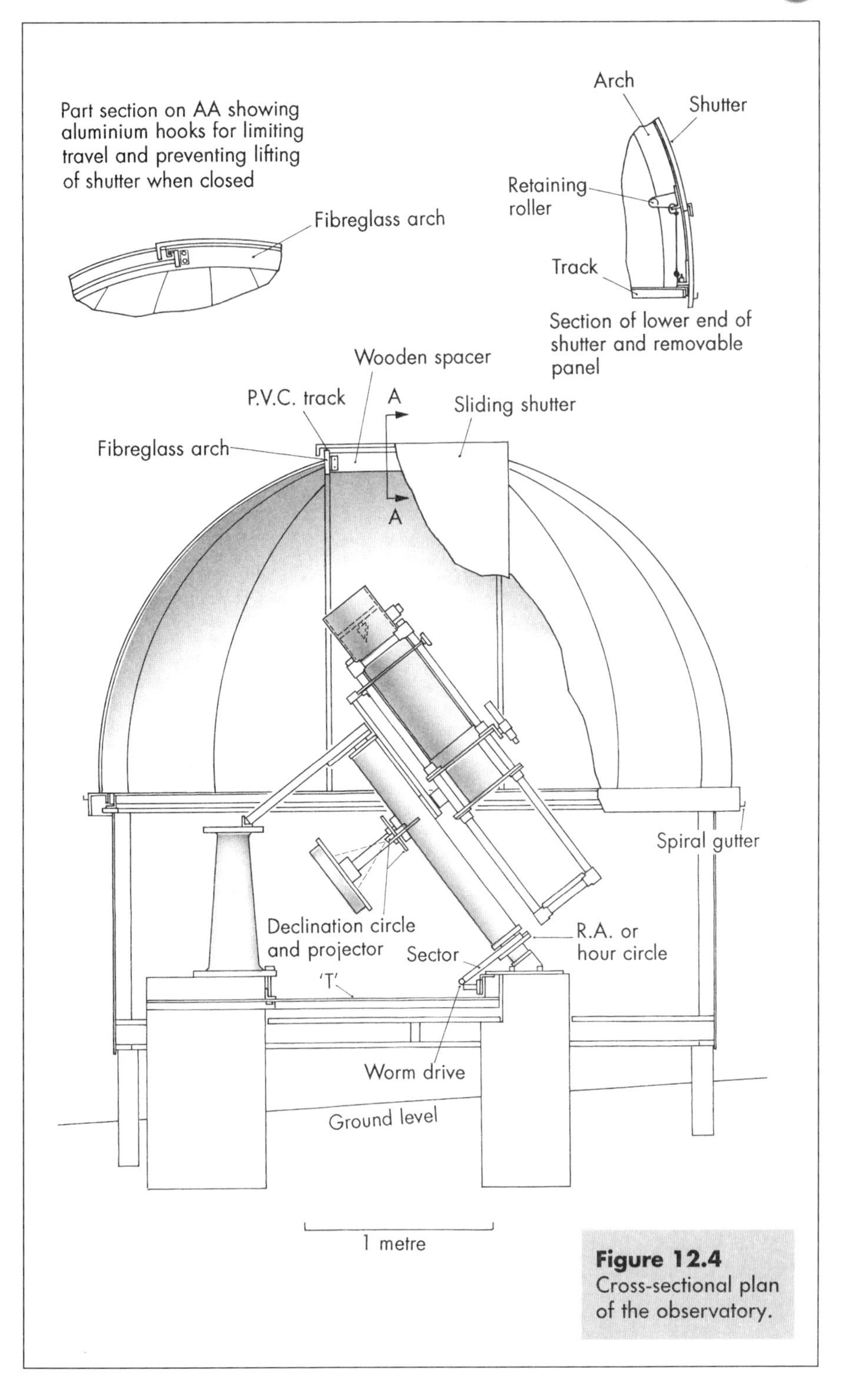

Figure 12.4 Cross-sectional plan of the observatory.

nels (or "gores"); one is a panel opposite the opening with its sliding shutter (see Figure 12.4).

The dome track was made in two halves, and was rolled professionally by a friend from 50 mm × 50 mm × 5 mm (2 in × 2 in × 3/16 in) aluminium alloy angle. Each half was made greater than a semi-circle, necessary because the rolling process means that a length at each end is always imperfect, and also because two 450 mm (18 in) lengths were required as bolting plates to make a rigid connection between the two halves (see Figure 12.4).

When finished the track was circular to within about 15 mm. Heat-treated aluminium alloy is springy and very difficult to roll; un-alloyed aluminium would have been easier although more prone to corrosion.

Making the Polystyrene Components

I was now ready to start cutting and sealing the polystyrene parts. Because this must be done in the dry, and each stage allowed time to dry thoroughly, I used my garage.

The first parts to be made were the arches, which are 25 mm × 115 mm (1 in × 4½ in) in section and cut from 25 mm (1 in) polystyrene sheet. I very carefully marked out six arcs, and cut the polystyrene to the correct radius and angle at the ends. I then pasted them and covered them with newspaper.

I used a simple wooden jig to hold pairs of these arcs while they were joined together, with paper pasted around the join and allowed to dry (see Figure 12.5).

The ends of the arches were then tied to make the exact diameter, and trimmed to the correct height. After sealing the ends, I covered the arches with a double thickness of fibreglass tissue and resin. Tissue is best because it is difficult to mould mat or fabric round a sharp corner. For extra strength a continuous band of 25 mm (1 in) glass fibre tape was added to the inner and outer surfaces of the arches.

The finished arches were later used as formers for the panels and sliding shutter while they were being

Figure 12.5 The wooden jig used to hold pairs of arcs while they were joined together.

sealed (see Figure 12.6). Later – after assembly – the outside of the dome was covered with a layer of woven glass fabric to give extra strength to protect it against gale damage.

The shape of the panels was calculated as described in Figure 12.2 and marked out on sheets of 12 mm ($\frac{1}{2}$ in) polystyrene. A flexible strip of wood was curved to join up the ordinates and a line drawn through them with a very soft pencil. The panels were then cut out using a sharp blade screwed to a block of wood. The blade was set to an angle of 45° to give a slicing action, and also inclined sideways at 12° to match the angle between the panels (Figure 12.7). It is important

Figure 12.6 The finished arches used as formers for the panels and sliding shutter.

Figure 12.7 The shape of the panels was marked out and cut from sheets of 12 mm polystyrene.

to remember that there are right-hand and left-hand panels!

After all the panels and the shutter and rear panel had been sealed, six of the panels forming one side of the dome were brought together and held with gummed paper strip. I found this operation was quite satisfying, as the panels fitted very well and in a very short time I had half a dome! Next, one of the arches was placed against a wall of the garage (which had been checked for squareness with the floor) and the assembled panels were fitted to it and fastened together with resin and tissue.

When the resin had set, the assembly was temporarily placed in position on the aluminium track and fibreglassed all over. This had to be done outdoors on a fine day, partly because of the size of the structure, and partly because good ventilation is essential because fumes are given off by the polyester resin.

Even with my wife helping me, it was only just possible to complete the outside of one half of the dome in one day. The inside was done on the following day.

The other half of the dome, the shutter and the rear panel were treated similarly. Then came the great day when my wife and I put the track (the base-ring) on the wheels on top of the observatory walls, and then lifted the dome sections and placed them in position.

The parts fitted nicely, so all that remained was to make the connections to the base-ring (see Figure 12.3) and to seal the rear section to the arches and place the shutter in position.

Later, L-shaped sections were made to form a skirt around the base to allow water to drip clear of the walls and to prevent rain being blown into the observatory (see Figures 12.1 and 12.3).

A further refinement, added later, was a spiral gutter (see Figure 12.4) which discharged water into a water butt for the garden. In total, I used 51 kg (112 lb) of R840 polyester resin supplied by BIP Speciality Resins.

I have to lubricate the shutter slides on the PVC sliding door track occasionally, with silicone furniture polish. The inverted T-section is fastened to the top of the arches, and the opposing inverted W-section, with the middle rib machined away, is fastened to the underside of the shutter.

The track was attached with polyester resin; the adhesion is not perfect but seems to be sufficient for the purpose.

The Building

The lower section of the observatory comprises twelve timber frames, each with a sheet of resin bonded plywood on the outside. The edges of the frames are made to an included angle of 30° so that they fit together without gaps. There is a wooden tongued-and-grooved floor, this and the walls having been most beautifully made by Mr Robins, our local carpenter.

The observatory is now about thirty years old and is still in quite good condition. It has never leaked rain, and only a tiny amount of wind-blown dry snow has got in on one or two occasions.

Looking Back

There are some things to be learned by experience.

Sealing the polystyrene sheet was a big job and polyurethane sheet – which would not need sealing –

might be better, particularly in an area where vandalism is a problem. A dome such as I have described is not resistant to damage from stones or air rifle pellets, and such damage would allow water to penetrate the outer skin, which would seriously damage the structure.

The laying up of the glass fibre is an even bigger job, one that should not be tackled lightly. Resin thickens enough to be unworkable in a few minutes so only small quantities can be mixed at a time. You will need plenty of solvent for brush cleaning, or an infinite supply of brushes!

Polyester resin, like most materials, is slowly degraded by sunlight. After some years, the outside of the dome grows a coat of algae and needs cleaning, then coating with more resin. More recently – and no longer having as much energy as I used to have – I just painted it.

When working on a dome of this type, remember that glass-reinforced plastic is brittle. My dome would *not* take a person's weight! When working on the outside, cleaning or painting, I use a short ladder with a cross board to spread the load – very carefully – and I am no heavyweight.

The plywood side panels were originally varnished with polyurethane varnish, but this again will not stand sunlight (no matter what make or type, marine or otherwise). It is not possible to revarnish over the top of the old coat, and getting it off a large area is a terrible job.

Having stripped all the original varnish, I have tried using one of the new microporous wood stains, which so far seems to be an improvement.

Finally, I think I should have made the observatory just a little bigger. When dressed up in a thick anorak against the cold, I find that another 300 mm (about 1 ft) on the diameter would have been useful.

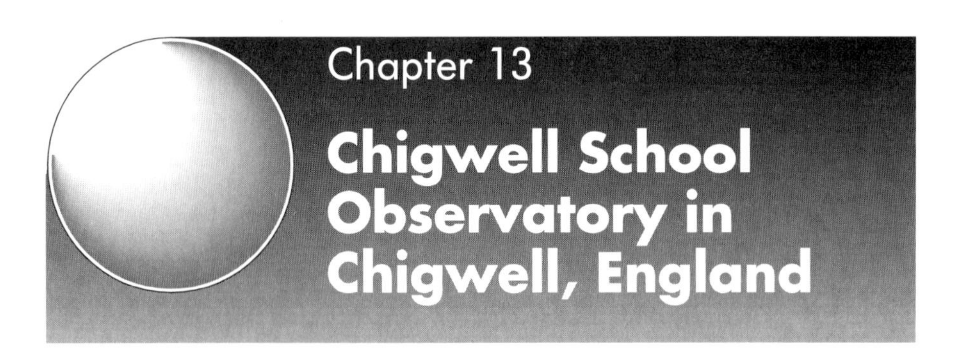

Chapter 13
Chigwell School Observatory in Chigwell, England

A.J. Sizer

The Replica Telescope

When I arrived at Chigwell School in the autumn of 1971, there was a thriving astronomy club in existence, but no observatory. There was a telescope, but this was rarely used as it was a curious beast. It had been constructed in the late nineteenth century by a master at the school and seemed to have been copied from one of the telescopes of William Herschel. It had an octagonal tube made of mahogany and was mounted on a wooden structure which was supported by four castors. The mirror was home-made of speculum metal!

As a museum piece it was magnificent but as a practical observational tool it was virtually useless. The mirror was tarnished and needed repolishing, but no astronomical dealer we approached was willing to tackle this task. The mounting was extremely wobbly even when it all stayed together and, because of the castors, it had a tendency to wander away on its own at vital moments. After one frustrating session chasing the thing around the car park while trying to observe Saturn, I suggested to the headmaster that we try to get something more practical.

In order to save the hassle of advertising, we wrote to Charles Frank, who at that time not only ran a telescope manufacturing concern in Glasgow, but also had a museum of antique astronomical equipment.

Figure 13.1 The second observatories: *left,* the 6 in observatory; *right,* the 10 in observatory.

He was delighted to accept our telescope and, in return, sent us a brand new equatorially mounted 6 in (150 mm) reflector and a Helios planetarium. It was with a certain regret that I watched the Herschel replica disappear out of the gate on the back of a lorry; I often wonder what became of it.

There was no doubt that the new telescope was a much more usable instrument than the old one had been, but there was still one problem which needed to be overcome. The telescope was stored in the physics laboratory. When anyone wished to use it he first had to gain access to the science building. This involved finding a member of staff who had a key; not an easy thing to do in the evening. He then had to carry the telescope down a flight of stairs and out on to the playing fields to find a site with a clear horizon. Having set the telescope up, he then carried out his observations. When he finished, he then had to carry the instrument back into the science building and back up the stairs to the physics laboratory. Small wonder that the telescope found little use: there were few pupils who possessed the required enthusiasm and energy both at the same time.

The First Observatories

I was having similar difficulties with my own 10 in (250 mm) reflector. I lived in a house in the school grounds and my telescope was stored in a garage. The 10 in was even heavier than the 6 in and needed to be carried out to the observing site in three pieces and reassembled on the spot. In the cold and dark this process took about half an hour and, after almost missing a total lunar eclipse when I dropped the vital bolt, I decided that something more convenient was needed. The answer was, of course, an observatory.

There was no problem getting permission for an observatory.

I think that our school is typical of many in the UK in that you will be allowed to build almost anything you like provided that (1) it doesn't get in anyone's way, (2) it doesn't cost anything and (3) you are prepared to build it yourself. But if a school doesn't have much money available, there is one thing that it possesses in abundance: cheap, enthusiastic (if unskilled) labour.

You would think that finding a site for an observatory in a school with fifty acres of fields would be easy. Not so. Many of these acres were taken up by football and cricket pitches and putting the observatory on any of these would obviously contravene condition (1). We wanted a site remote from any lights but not too far from the buildings: vandalism was an obvious worry. Eventually we settled on a place about a hundred yards from my rooms, in the corner of an old cabbage patch round the back of a boarding house.

When the school had been smaller it had grown most of its own food. As the school expanded in the 1960s this became impossible: a catering firm was engaged and the vegetable garden had become disused and overgrown. On first inspection, the site did not look too promising, resembling as it did a jungle of brambles and nettles. However, a session with a Hayter and a few bonfires later, it looked more suitable and, to the south at least, had a good horizon and few lights. Construction of the two observatories began in the summer holiday in 1976 and was finished by the start of the autumn term.

I had no experience then in building observatories. With conditions (2) and (3) in mind, it was obvious that nothing too ambitious could be considered. Certainly, any idea of a brick-built structure with a rotating dome was out of the question. In the end, it was decided to use wooden structures, just high enough to enclose the telescopes, with removable roofs and sides which would fold down to allow the telescopes, which were on low-slung German equatorial mountings, to reach objects near the horizon. The phrase "telescope hutches" entered into the vocabulary of the Chigwell pupils.

The design was by no means ideal. The roofs had to be removed in order to use the telescopes. They were heavy and awkward to lift, and even more difficult to replace than to remove. Because the observatories had been built as small as possible to keep the weight of the roofs to a minimum, there were certain positions in the sky which were inaccessible as the telescope would foul some part of the observatory structure. But, all things considered, the observatories were an improvement on what had gone before. With the added convenience, use of the telescopes increased considerably. It was even possible to invite pupils from local schools to observing sessions. Before such a session it was usually necessary to do some weeding to remove the vicious nettles, which thwarted all attempts to exterminate them and, if there were any girls in the party, at least the biggest of the spiders had to be persuaded to move elsewhere.

It was the increasing affluence of the pupils during the 1970s which brought about the demise of the first observatories. At this time, an increasing number of sixth-form pupils were driving to school in their own cars, leading to a need for increased parking space.

In 1981, it was decided to build a new car park to satisfy this need and the site chosen was the cabbage patch on whose edge stood the observatories. As a car park, complete with lighting, is a far from ideal place for observatories, it was obvious that they would have to be moved. Unfortunately, this eventuality had never been considered when they were designed and, in the attempt, they fell to pieces.

Replacement and Resiting

At the time there was little money available for replacements. However, in 1979 an old building had been pulled down and the wood of its floorboards had been stored. This was made available for our use. But, as the wood had been stored in the open, it was rather wet: some of it indeed was waterlogged. Yet, when there is no alternative, you take what you can get!

The new site chosen was at the edge of one of the football fields, next to the groundsman's house. With the experience of the problems of the previous design, it was decided to build the new observatories with roll-off roofs. As the roofs no longer needed to be lifted off, the observatories could be built considerably larger and so the telescopes would be able to reach all parts of the sky.

Construction began in the summer of 1982. It was not intended that the observatories would be permanent structures (with the wood available, this would not be possible), so there was no need to lay any foundations. This proved to be just as well because, just before the completion of the first structure, the school decided to illuminate the path from the school to the girls' boarding house with lamps which made no attempt to minimise light pollution. As the path passed within twenty yards of the observatory site, something obviously needed to be done. The action chosen was to transport the almost completed observatory on a trailer pulled by the school's tractor, a distance of about three hundred yards to a new site remote from any lights. I think that this must be an event unique in astronomical history!

By the summer of 1983 the two new observatories were complete. The roofs rolled off on roller skate wheels and, although inclined to be rather hard to push off, were a definite improvement on the lift-off roofs of the first observatories. Mains electricity was installed to run the telescope drives and any auxiliary equipment that needed it. In October 1984 the school played host to a number of eminent astronomers who were invited along to perform the opening ceremony.

For a few years the telescopes were used with greater ease than ever before. But, by 1988 it was obvious that the 10 in reflector needed a new mount.

Figure 13.2 The third 10 in observatory and the 10 in reflector.

The original one, built in 1969, was good enough for visual work but was not sufficiently accurate for long-exposure photography.

The telescope was dismantled and sent away for reconstruction. A Celestron C8 was purchased as a replacement instrument but was too high to be accommodated in the large observatory: the small observatory was modified to take it and the 6 in removed altogether. As the 6 in had no motor drive, it had seen limited use in recent years. It was decided to convert it into a solar telescope by removing the aluminium from the mirrors, a task which was successfully completed. As a temporary measure, it was parked in the school porch while a more permanent home was a found for it: unfortunately it was stolen before this could be done!

In 1989 the 10 in telescope returned on its new mounting which proved to be a marked improvement on its predecessor. The C8 was sold and replaced with a Vixen SP102F fluorite refractor on a tripod mounting which allowed it to be kept indoors and used as a portable telescope and also, with the aid of an Inconel full-aperture filter, as a solar telescope. Later an H-alpha filter was purchased which allowed observations of flares and prominences.

By now the temporary nature of the observatory was becoming apparent. The wood, never very sound, was beginning to rot badly. Vandals broke in one night but, although they further weakened the structure of the building, found nothing inside to their liking and contented themselves with writing what were presumably rude words on the wall with what they thought was spray paint but which was, in fact, the can of compressed gas which had been used to blow dust off the corrector plate of the C8.

Rabbits set up home beneath the floor of the observatory and began their habitual excavations, leaving great piles of earth all around the telescope. It was clearly only a matter of time before there was a serious accident, either due to someone falling down a burrow in the dark or due to collapse of some vital part of the observatory structure.

By this time the school authorities had appreciated the benefit to the school of an active observatory. It was an asset that appealed to parents of prospective pupils and was useful to our own pupils when they sat the GCSE astronomy examination. When I told them that the building needed to be replaced before it fell down, they allotted some funds for a new one. Flushed with this unaccustomed wealth and tired of periodically building observatories, I decide that, this time, we would do it properly.

The New Observatory

One of our pupils lived in a large house nearby. His father loaned us the services of his estate manager and some of his labourers for the duration of the project. In the summer of 1993, work began. The site was cleared and a concrete foundation laid. The rabbits disappeared, apparently unhappy about living on a building site.

Brick walls, supporting the wooden runners for the roll-off roof were constructed. A central area of the floor was insulated from the surrounding concrete with polystyrene foam: this would be the base upon which the short tripod for the mounting would sit: the polystyrene would prevent vibrations from the floor being transmitted to the telescope. The wooden roof was supported on ten industrial castors of tough nylon running in aluminium channels, making it very easy to roll off when required. A carpet was laid to keep down the dust and to prevent air currents, rising from the concrete floor, disturbing seeing conditions. Cupboards were built to take auxiliary equipment which was planned. By the autumn, the telescope was in place and ready for action.

It was immediately apparent that the new observatory made use of the telescope very much more convenient than it had ever been before. It was now possible to do some useful work even when only an hour was available. Indeed, one evening I walked down to the observatory, took some photographs of Jupiter and, when clouds rolled in, closed up and returned to my room less than half an hour after leaving it! Early morning observations of the waning crescent Moon could be made without getting up in the middle of the night.

Over the Christmas of 1993 and the spring of 1994 I built a Starlight Xpress CCD camera from a kit of parts. It took a while to get it working and to obtain a suitable computer to work with it but soon I was able to take some images of the Moon and Mars. It became clear, though, that improvements were necessary to the drive of the 10 in before full use could be made of the camera: because of its very small field of view fine slow motions on both axes would be required to allow objects to be centred.

In the end, we decided to completely remount the telescope on a much sturdier and steadier mount: by the winter of 1994–5 this had been done and the instrument was again ready for action. Unfortunately, the spring of 1995 was exceptionally cloudy and the opportunities for using the telescope proved to be few. This period was also very wet and a moat had to be dug around the observatory to prevent flooding! During the autumn, the computer and camera were installed in their cupboards and I waited in vain for a clear evening which was also free of the need for marking or report writing. On two occasions, pupils from neighbouring schools came around in the hope of seeing the Moon through the telescope but were unable to do so: instead they saw pictures of it taken through the telescope. We hope for better luck in the near future!

Location

Chigwell is situated to the north-east of London, about fifteen miles from the centre. Although it is right on the outskirts of the built-up area, it is badly affected by streetlights to the south and the sky is far from dark. For this reason, our main field of interest is the Moon and planets. From the observatory, there are good southern, eastern and northern horizons, though the sky to the west is obscured by trees to an altitude of about ten degrees. There are no lights within about three hundred yards.

Instruments

The 10 in telescope has a primary mirror of focal length 80 in and was made by J. S. Hindle in 1963 to a claimed accuracy of one-twentieth of a wavelength. There are two interchangeable Newtonian secondaries, one (for planetary work) made to the minimum size necessary to ensure complete illumination of the centre of the field of view: the other rather larger for wide-field work. In practice, only the first of these is used as it seems perfectly satisfactory for all purposes. Focusing is achieved by sliding the eyepiece mount, complete with secondary holder, up and down the tube, thus ensuring the minimum distance between the secondary and focal plane: this is necessary in view of the small size of the secondary mirror.

Objects are picked up with a Telerad, then centred using a 7 × 50 finder. There is a 50 mm (2 in) long-focus refractor for guiding purposes. Objects can be viewed in the normal way with an eyepiece, or with an image intensifier unit. A camera can be attached in prime-focus or eyepiece projection modes, or can be used with the intensifier. It was with the intensifier that I obtained my first picture of M57, the Ring Nebula. It showed the central star clearly: exposure time was 1 second!

The CCD camera can be attached in a few seconds. It can take images using exposure times between about 1/1000 of a second and several minutes: images are transferred to an IBM 286 PC. From there they are taken (on floppy disks) to an Acorn A5000 for processing.

A small video camera module can also be attached directly to the telescope. This is sufficiently sensitive to allow real-time pictures of the Moon and planets to be displayed on a monitor screen, where they can be viewed by several people at the same time: very useful when a group of visiting pupils come around.

The 10 in is, of course, much too large to be used as a solar telescope. For this purpose the 4 in Fluorite is used. Inconel filters enable the Sun to be viewed and photographed and prominences can be observed with the hydrogen alpha filter. The fluorite can also be used to observe objects that the 10 in cannot reach, for instance Mercury in the evening sky, which is obscured by trees.

Various small telescopes are available for the use of students who are studying for the GCSE astronomy examination. These include a 6 in (150 mm) reflector and a $3\frac{3}{4}$ inch (96 mm) refractor, both home made from parts obtained from government surplus stores.

Our next project is the construction of a planetarium for use both by the school and by visitors. The projector has already been obtained: all that we now need is sufficient money to allow the erection of the building to put it in. I don't think that my inexperienced bricklaying is up to this task!

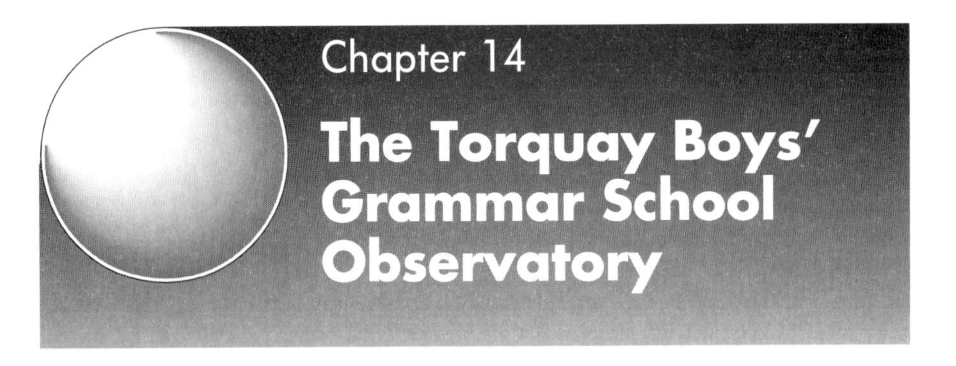

Chapter 14

The Torquay Boys' Grammar School Observatory

David Reid and C. Lintott

The Torquay Boys' Grammar School Observatory is part of Torquay Boys' Grammar School in Devon. Torquay lies in a small, east-facing bay in the south-west of England, at about 50.45° north, and 3.5° west of the Greenwich meridian. The observatory is well inland, situated close to the Devon countryside. Its two main buildings are on ground that is sloping upwards away from the school and so we have a clear view over the top.

Figure 14.1 The Observatory at Torquay Boys' Grammar School

There are two parts to the observatory; a dome section and an adjoining workroom/control room that have been built facing along a north-south line. There are short banks on three sides so that the observatory is sunk down slightly (see Figure 14.1). It houses a 0.5 m ($19\frac{1}{4}$ in) Newtonian reflector with a fork-type mounting and a fully computerised control system which is used for visual, photographic and CCD astronomy. This range of equipment is used for a variety of work, from education (both adults and children) to detailed research work.

We opened the observatory in October of 1987, but it was at the end of 1985 that the first ideas for it formed. One or two of the teachers at the school along with the local astronomical society considered a moderate telescope, of around 15 in (380 mm) aperture, in a simple, inexpensive, dome. However, we found, through Broadhurst, Clarkson and Fuller Ltd. in London, a part-completed 0.5 m ($19\frac{1}{4}$ in) reflector and so we had to reconsider things.

The observatory would have to be larger than originally thought, and a series of successful fund-raising events made the current setup possible.

The buildings were designed with the telescope in mind and we had the help of the late J. Hedley Robinson, FRAS, who was a patron of the Torbay Astronomical Society at the time. We drew up the basic designs ourselves, then found an architect who was happy to draw up full plans for the builders to use. All in all the project cost in the region of £25,000, and was built in a matter of months – the buildings being finished in a summer (around three months).

The Building

The shape is quite classical; a rectangular room with a circular dome part attached. This base room is 3.58 m (13 ft 8 in) in length and 2.78 m (9 ft 2 in) wide. The foundations are similar to those of a normal building (we had trenches dug which were later filled with concrete). The walls of the workroom and dome section are two-layered and there is a gap, or air cavity, between the two. In total the cavity walls are 250 mm ($9\frac{3}{4}$ in) thick.

The inner bricks are breezeblock and the outer ones small decorative red bricks which match the existing school buildings. The roofing is relatively normal for any building; a flat roof consisting of timber joists topped with 25 mm (1 in) thick felt, asphalt and chippings to finish. The ceiling is plasterboard and plaster, nailed and painted afterwards. This roof does have mineral wool insulation between the timber strips, which is not always a good idea in observatories.

Bearing in mind that the observatory points north-south, the door lies on the east-west line. Along this east-west wall there is a sink unit, with full plumbing to one side of the door (which opens inwards) and a table top, with cupboard space below, to the other. This is where we store our photographic enlarger. One large benefit of this workroom not having any windows is that it can double as a darkroom, a useful facility. Along both the north and south walls are high cupboard units similar to those used in kitchens, and low floor cupboards topped with workbenches or desks. Two small strip-heaters are fixed on opposite walls, to heat up the room on cold winter nights. Although heat rising across an observer's view is infuriating and misleading when making an observation, these heaters are particularly useful if it clouds over or if work is being done during the day. On both sides, near the door leading to the dome, are specially designed computer desks: one for a PC, the other for the computer – along with its monitor – which controls the main telescope.

Connecting the base room with the dome is a simple timber door leading to steps curving slowly upwards 1.05 m (3 ft 6 in). In other words, the floor of the dome is 1.05 m above the workroom floor. The original designs for the observatory actually show a double door system, which would have been a much better, though awkward, idea. There would have been double doors, then a space of about 0.5 m (1 ft 6 in) and another set of double doors. The idea being that one would shut the first set of doors before opening the second and going up into the dome. This would have created a kind of air lock, preventing warm air circulating up into the dome every time the door was opened. But with having to fetch equipment, and continually using the downstairs computers to aid observations, it would have been decidedly cumbersome.

The Dome

The dome itself is made from galvanised steel, which started life in the form of a grain silo top! It was imported from America by Alan Young, who designed and made the dome for us. It originally arrived on site in two half-hemispheres which were later fixed together and lowered into place by a crane which we had borrowed from a local firm (see Figure 14.2).

The dome rests upon sixteen nylon wheels which allow it to rotate easily. In order to do this a ring, or strip, of metal teeth is attached to the dome about 355 mm (14 in) from the base; a large cog interlocks with this and is turned by a 415-volt three-phase motor. The motors and metal strip are clearly shown in Figure 14.3.

Making this mechanism must have been an incredibly time-consuming job, as much of it was done by hand. For example, every tooth on the metal strip was individually cut and welded on (there are over 300 of them going the full 360° around the dome)! The Farvalux motor from Bournemouth in England is connected to the three-phase supply via a control box with simple east-west functions.

Figure 14.2 The dome being lowered into position.

An alternative design that was considered was a strip of metal teeth fixed to the wall, and the motor attached to the metal dome (reversing the situation). However, I think that a stationary motor is by far a more effective solution.

Another motor of similar type aids the opening and closing of the slit of the dome. The shutter – a metal sheet which rides backwards and forwards over the dome – is connected to ropes which are pulled taught and released in turn by a grooved cylinder, or worm (see Figure 14.3). The single rope runs out of the worm on one side, up along the inside of the dome, and is attached to the base of the movable slit. This is pulled back by the motor in order to raise the slit. When the slit is raised, the other end of the rope, which comes out of the opposite side of the worm, slackens – it is attached in a downwards-pulling position (better explained by looking at Figure 14.4) and so allows the slit to ride up and back. When the slit is closed the effect is reversed, the downwards-positioned rope-end pulls down while the other is loosened.

Figure 14.3 The "worm" which operates the ropes to open and close the dome slit.

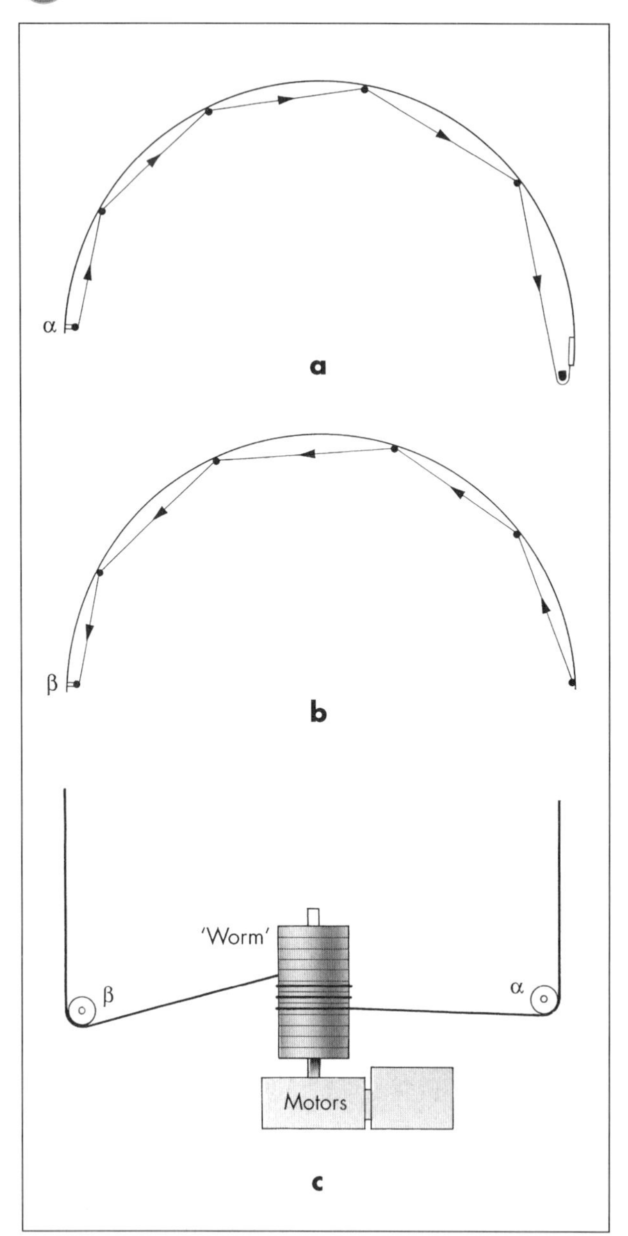

Figure 14.4 The means by which the worm opens the dome slit. **a** Cross-sectional view of slit. **b** cross-sectional view of slit from the other side. **c** front view of worm and ropes seen in **a** and **b**.

The Telescope Plinth

The radius of the hemispherical dome, from the top to the centre where the telescope lies, is 2.133 m (7 ft 9 in). The radius of the circular building that the dome rests upon is 1.965 m (6 ft 6 in), again with the telescope almost dead centre. As you can see from Figure 14.5 the interior of the steel dome is entirely covered with polystyrene tiles; these help to combat condensation. Around the circular walls of the dome section are the dome control boxes already mentioned, and four white lights as well as four large red lights, invaluable in allowing you to retain your night vision while still being able to see what you are doing.

With a telescope as large as a 0.5 m reflector an important consideration is the base and the ground upon which the telescope will stand. As is the case with most large telescopes, we have a large concrete block which is often referred to as the plinth. The plinth we installed is 4 ft × 5 ft × 6 ft deep (1.2 m × 1.5 m × 1.8 m) and can be seen in Figure 14.6. It is rectangular, with a raised triangular top. We lined the 6 ft hole with expanded polystyrene before the concrete for the block was poured. The polystyrene acts as a kind of shock absorber, isolating the plinth from the surrounding floor. The floor here is covered with a thin carpet, while the downstairs floor is fitted with linoleum, which is far easier to keep clean.

Figure 14.5 The open dome, with polystyrene tiles covering its interior surface.

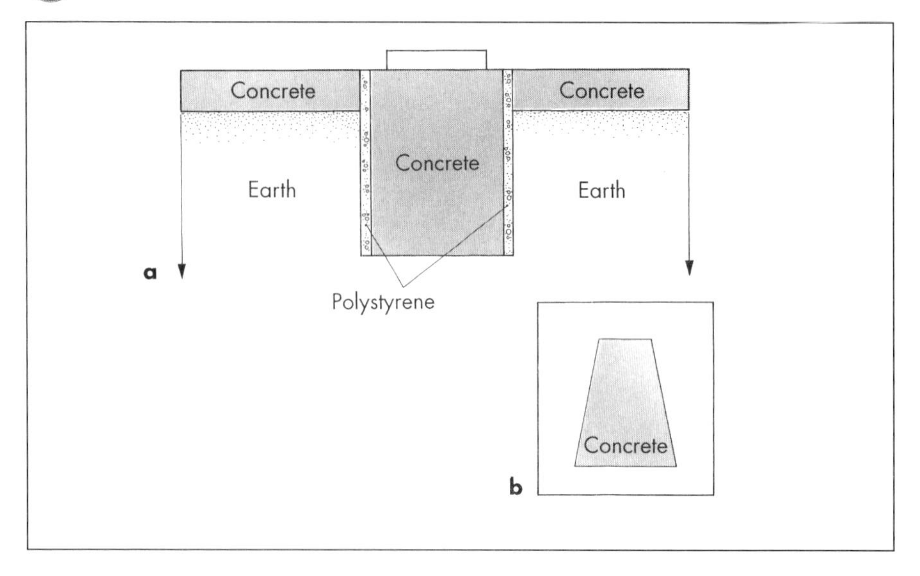

Figure 14.6 The plinth used to support the 0.5 m refractor is 4 ft × 5 ft × 6 ft deep. **a** cross-section. **b** plan.

A Few Problems, but an Overall Success

Something that we didn't consider to be a significant problem during construction was drainage. And so we put in a small square drain on the north-south facing wall, near the door. Water from the guttering ran down a vertical pipe into this drain. It proved to be completely inadequate, and the result has been a significant amount of damage.

As I've already said, the buildings are sited on an upwards-sloping enclosure, leading up to rugby and cricket fields above and beyond the observatory. The amount of surface run-off and ground saturation that we get here in the southwest is considerable: the local weather is very wet indeed.

Rain has caused one or two floods in the winter, the drain being easily blocked by leaves. The dome was of course unaffected, but the workroom below filled with about 7 in (180 mm) of water. We have now remedied the situation, by hiring a small digger and excavating a much larger and more effective drain, down towards the school's central water collection points. This is therefore an important consideration when building a permanent structure of this sort.

As I have already mentioned, everything is sunk down below the normal level of the land. This was an attempt to make the observatory as inconspicuous as possible, an important consideration in modern times when property is at such a high risk of vandalism. The school grounds are easily accessible to the public and we were in a predictably vulnerable position. Even though we have had problems with graffiti, with hindsight, we should have made the building considerably higher. This would have allowed us an improved view in some directions, and is one of the few things we would have done differently if starting from the beginning again.

However, all projects of this type have a few problems. Overall, the Torquay Boys' Grammar School Observatory has been a complete success and has made quite an impression upon the community. The equipment has performed at high levels, allowing research to be done in comfort.

The observatory has been host to many visits and is an invaluable educational tool in the southwest. The buildings have been used as laboratories, workshops, darkrooms, and computing posts during the day and evening. Solar work has been done along with deep-sky and planetary work. The observatory has performed even better than was hoped, becoming larger and more effective in its local role, and has succeeded in educating a community of adults and school children alike, in astronomy.

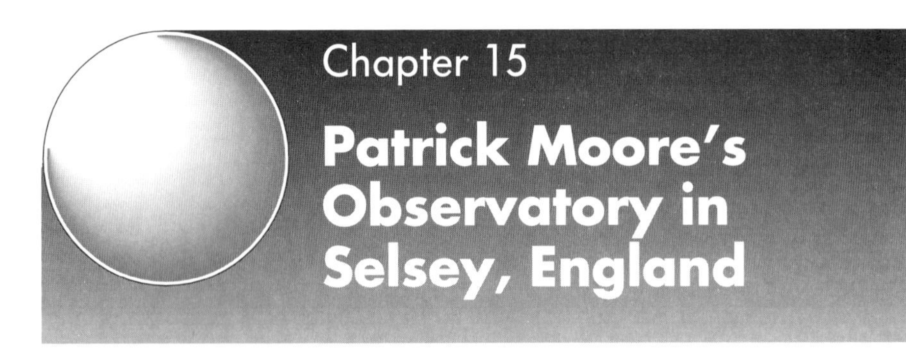

Chapter 15

Patrick Moore's Observatory in Selsey, England

Patrick Moore

As I am essentially an observer of the Moon and planets, it may be said that my observatory is of the "old-fashioned" type, and this is no doubt true enough! There are four main telescopes: 15 in (380 mm), 12.5 in (317 mm) and 8.5 in (216 mm) reflectors, and a 5 in (127 mm) refractor.

The 15 in reflector has a wooden octagonal tube, partly enclosed, and is on a massive fork mounting; there is a revolving head, so that the eyepiece can always be kept in a convenient position, and there are

Figure 15.1 The observatory housing Patrick Moore's 15 in reflector.

Figure 15.2 The top section of the observatory is easily moved by means of its toothed inner ring operated by turning the handle.

three finders. It has a normal electric drive, with electric slow motions, but the tube can be moved by hand even when the drive is running, which is always helpful.

The observatory "dome" looks rather like an oil drum (see Figure 15.1). It was made roomy because there are occasions when television crews are using the telescope, and the entire top section of the building moves round. This is managed by use of a toothed inner circle, and the whole section is extremely easy to move merely by turning a handle (see Figure 15.2).

There is a window in the upper section; this is opened first, and then two sections of the roof can be swung back, by means of a handle, on the supports (see Figure 15.3). The dome is asymmetrical, so that it

Figure 15.3 The open window and roof section.

Figure 15.4 The run-off roof shed with its roof slid back.

is possible to reach the zenith. The dome itself is not driven round, but a slight adjustment every half-hour or so is all that is needed.

There is not a great deal of light pollution (there is sea on three sides of the observatory), and there is only one inconvenient tree which is on adjacent land and which, unfortunately, I have been unable to prune. A second tree, on my ground, once produced a modest crop of pears. One night it obstructed the view of Saturn, and the next day it turned into a small, stumpy tree – which produces many more pears than formerly!

Figure 15.5 The run-off roof shed and the 15 in dome.

Close by the 15 in observatory is the run-off roof "shed" housing the 5 in refractor (see Figures 15.4 and 15.5). The telescope is an excellent one (the object-glass is a Cooke triplet) and it has been set on a conventional pillar mounting, with electric drive (see Figure 15.6). The original was made of plastic with wooden supports, but has now been replaced with an all-wooden construction. The roof is moved back by means of a chain arrangement – in fact, old cycle chains were used – and when the telescope is to be used, the top sliding roof is supported on an extension.

The third telescope, the $8\frac{1}{2}$ in reflector, has a With mirror and a Browning mount, on the German pattern with a massive counterweight. The top section moves round on a rail, and is easy to turn by hand.

Figure 15.6 The 5-in refractor housed in the run-off roof shed.

The observatory, originally set up at my old home in East Grinstead, was made to look "decorative" (see Figure 15.7) because in that site the only place for it was in the middle of the front lawn. Certainly it is no eyesore, and it is effective, but it has two disadvantages. First, the glass windows mean that the inside temperature can rocket, and one has to "open up" well before starting to observe. Secondly, entry has to be via the lower section, and means crouching down.

The $12\frac{1}{2}$ in reflector is on an altazimuth mounting (see Figure 15.8). This has the obvious disadvantage that it has to be hand-guided all the time, with manual slow motions, and it cannot easily be used for photography. On the other hand it is convenient and simple, and for my limited amount of variable-star work the telescope can be swung very quickly from one side of the sky to the other.

The run-off shed is in two parts, and runs on rails (see Figures 15.9 and 15.10). The two halves are pushed back in opposite directions, and do not obstruct the view of the sky; a two-piece shed of this type is far better than a single-shed arrangement. With a single shed, there must be a door. If hinged, the door flaps. If it is removable, there are problems in replacing it on a dark, windy night; the door tends to act in the manner of a powerful sail.

Figure 15.7 The "decorative" observatory housing the $8\frac{1}{2}$-in reflector.

Figure 15.8 The $12\frac{1}{2}$-in reflector on its altazimuth mounting.

The main drawback of a run-off shed of this kind is that it gives no protection against the wind force during observing, and there are also artificial lights to be considered. In my own case there was only one inconvenient street light when I came to Selsey (I persuaded the local council of the time not to put another one on my hedge!) and I have screened this, as shown in Figure 15.10. Also shown are the observing steps for use when the telescope is pointing at high altitude, and a table to hold eyepieces and other materials.

All in all, the observatory suits me well; there are no "high-tech" computers and electronic devices, though no doubt these could be added if need be.

A final word of warning. The $8\frac{1}{2}$ inch dome was for a time at Armagh in Northern Ireland (I was Director

Figure 15.10 (*opposite*) The closed run-off shed, with (*in background*) the screens used to shield an inconvenient street light.

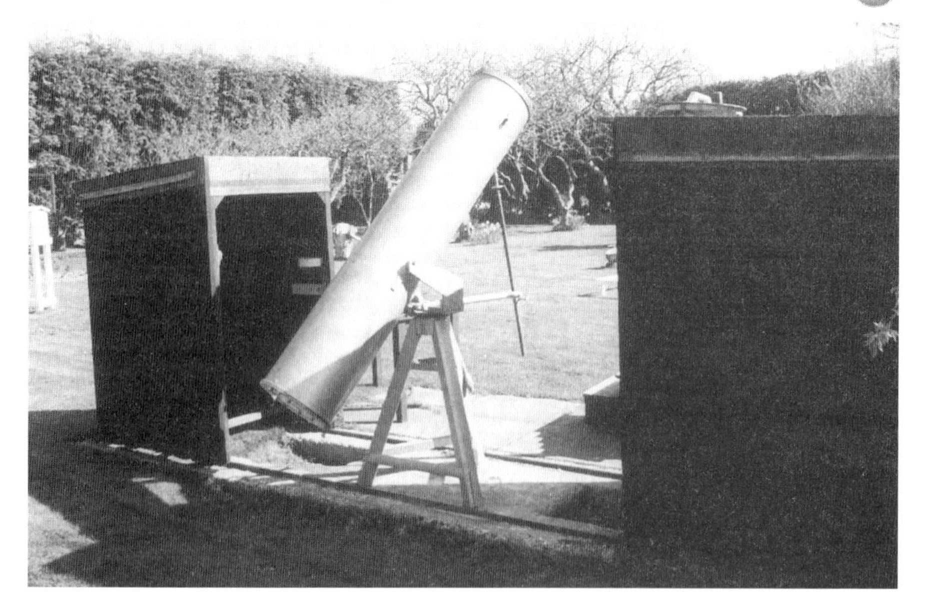

Figure 15.9
(*above*) The two-part run-off shed provides an unhindered view of the sky.

of the Armagh Planetarium from 1965 to 1968). When I moved back to England, I sold my house. The purchaser suddenly claimed the dome, on the basis that it stood in the garden. In fact the claim was invalid, because the dome merely rested on a concrete base and was not fastened down (it is so heavy that fixing it is unnecessary), but I did not wait; within

hours the dome was not only dismantled, but on its way to Selsey. However, do bear in mind that if you sell your house, you should make sure that any astronomical equipment is protected against any last-minute claim.

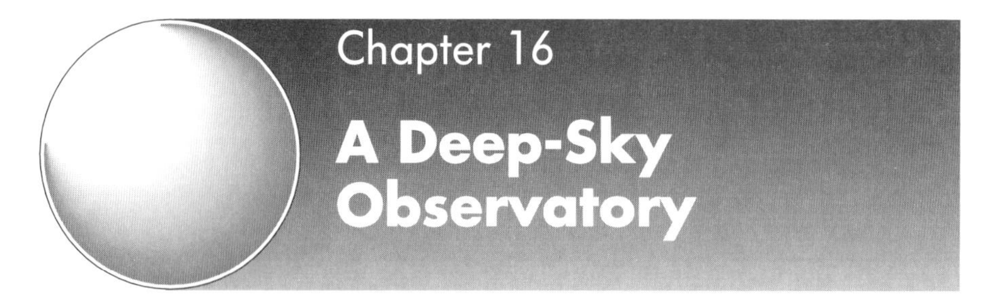

Chapter 16
A Deep-Sky Observatory

Jack Newton

I have been an amateur astronomer for almost forty years, yet today I find the universe as mysterious and intriguing as I did during my childhood. I took my first astrophotographs at age twelve, in an attempt to prove to my school chums that I really *could* resolve the rings around Saturn, even though my instrument was a tiny refractor and my vantage point the roof of a parish church! Since that time, I have constructed a number of observatories in locations all across Canada. Whenever I found myself "between properties" or fleeing light pollution, I ventured forth with mobile

Figure 16.1 The Newtons' home in British Columbia, housing the observatory, darkroom, machine shop, computer room and home cinema.

telescopes. What follows is a description of the "observatory with living quarters attached", which I have been enjoying for the past four years.

Before beginning, I should explain that I never quite got the hang of reading blueprints or following building plans. All of my previous observatory designs, stationary and mobile, reflect my personal belief that function takes precedence over form. If something isn't available off the shelf, then I design and build my own. If something doesn't fit, then I get a bigger hammer! My usual strategy is to watch the "doers" and then adopt their most noteworthy ideas. I'm unimpressed with expensive, flashy setups which look great but never get used. I unreservedly "borrow" the best ideas I see in other observatories, and am delighted if other people choose to adopt mine. I devote a great deal of time to astrophotography; however, I am devoted to public education in astronomy, and frequently open my observatory to students of all ages. I get a lot of enjoyment out of teaching young people visual observing techniques, and introducing them to the wonders of the universe through computer imaging.

I realised my dream of one day building a mountain-top home, with an observatory on the roof and a large telescope inside when my wife, Alice, and I discovered $7\frac{1}{2}$ acres of property on the extreme south-western tip of Vancouver Island, British Columbia, Canada. The heavily treed land overlooks the Straits of Juan de Fuca and the beautiful Olympic Mountains in Washington State, USA. This site is on Mount Matheson, 300 m (1000 ft) above sea level. It is not a lofty height as far as mountain elevations are concerned, yet is usually sufficient to keep the fog and ground haze well below us.

I trailered my telescope to an adjoining parkland site for almost three years before committing to the mountain location. I did all I could to test the suitability of the site beforehand. Anyone who has ever tried to forecast weather for coastal areas will confirm that such predictions constitute a far from exacting science. Conditions tend to vary quite considerably hour-to-hour, much less over days and weeks. Generally, the local weather from January through June seems to follow no set "pattern". And decent seeing during July and August can be disturbed by winds down the Strait. As often as not, though, September through December make up for the other months of uncertainty, and frequently offer still air with arc-second seeing.

Alice and I self-designed our new home (see Figure 16.1) to include all the favourite features of past homes in which we'd lived and allowed for the addition of a few extras: the observatory, an 11-tonne telescope pier, a photographic darkroom, a machine shop, computer room and home cinema. My warm office is in the level directly below the observatory, and from here I can manually or auto-guide. We planned the house in such a way as to allow plenty of room for the pursuit of our many varied interests. I knew from previous experience that I would seldom have a night alone in the dome, and so paid particularly close attention to details enhancing the comfort, convenience and safety of the observatory. We included the theatre not only for our own enjoyment, but also as a projection room where we can entertain the visitors who overflow the dome during observing sessions and tours.

I designed and built the 4.8 m (16 ft) dome myself, after regular work hours and on weekends. I was up against a rather gruelling time schedule of thirty days from start to finish. This was due to a number of factors. First, the unexpectedly fast sale of our previous home (which we needed to sell in order to free up money for the new one) had forced us into rental accommodation on only three weeks' notice. Second, the new construction progressed quickly, and any delay in the dome would cost our building contractor lost time (and us more money). So the dome had to be completed, delivered and ready for lifting by crane on the same day that the roof trusses were put into place. Finally, just to add to the excitement, we were ticketed to leave for the solar eclipse in Baja, California only a day or two after the scheduled lift. There was no margin for relaxation!

I did the fabricating in a friend's spacious and well-equipped workshop some distance from our home. I needed to construct the dome in halves, so that it could be more easily transported to the new site.

I started by cutting curved 200 mm (8 in) widths of 30 mm ($\frac{3}{4}$ in) plywood from 1.2 m × 2.4 m (4 ft × 8 ft) sheets and positioning them on the shop floor to form a 4.8 m (16 ft) circular base. I then glued and screwed two additional plywood layers over the first one to create a strong laminated base-ring. I built the two main overhead arches using the same diameter as for the base and set them 1.2 m (4 ft) apart. I then laminated strips of 4.8 m (16 ft) × 60 mm ($2\frac{1}{2}$ in) × 10 mm ($\frac{3}{8}$ in) cedar lathe (the kind commonly sold by garden shops to make rose trellises) to form a 100 mm (4 in) thickness and bent these into position, C-clamping them to the circular base as a temporary bending form. I needed twenty-two of these laminated ribs, which I had also glued and screwed for strength. I then formed the front of the dome by placing the ribs at 0.6 m (2 ft) intervals from the outside of the base-ring to the arches (see Figure 16.2). I secured them into place using metal brackets. I then covered the "skeleton" with 3 mm ($\frac{1}{8}$ in) thick mahogany door-skins (see Figure 16.3). This sheeting is the thin material used by manufacturers to cover hollow (as opposed to solid) core doors, and is usually available from a builders' supply store in 1.2 m × 2.4 m (4 ft × 8 ft) sheets. I shaped and cut each section, bent it over the rib structure, glued it into position over the ribs, and then nailed it into place. Once all of the panels were in place, I painted the interior of the dome black to cut down on reflected light and sprayed it with Lysol disinfectant as a mildew retardant. Finally, I covered the outside of the dome with fibreglass cloth and topped it with two coats of resin and a final white coat. Once dry, the dome was finally ready to be moved out of the workshop!

Figure 16.2 The overhead arches, laminated ribs and mahogany sheeting used to form the dome structure above the base ring.

Figure 16.3 The mahogany sheeting seen from the outside.

Unfortunately, the only egress from the workshop was down a steep, narrow, twisting and gravelled access road and then along a public roadway for several miles. We made this move after nightfall, because although the dome halves were neatly sandwiched back-to-back for transport, their width still exceeded the legal limit for open-road hauling without a special permit.

Save for the weather, which was cold, wet and windy, the move to the new neighbourhood went without a hitch. We celebrated our success with a champagne toast, hunched inside the dome alongside the roadway. Somehow, I couldn't help wondering if this experience bore any resemblance to a party in an Arctic igloo. . . .

The next day, I finished assembling the dome by reconnecting the halves of the base. My final crucial step was to bend electrical conduit to form a ring, which I attached to the bottom of the dome.

I then turned my attention to the portion of the house roof where the dome would sit. I had a 1 m ($3\frac{1}{2}$ ft) pony wall built on the roof, and I installed five in grooved V-wheels in an upward position every 0.3 m (1 ft) along the top of this wall. The idea is that when the dome is lowered, the conduit ring on the bottom of the dome becomes the track in which the wheels ride. I'm pleased to say that my prayers were

Figure 16.4 The dome about to be hoisted into position on the roof.

ultimately answered as the dome was lowered over the track, the two circles matched up perfectly and the dome rotated freely. With the dome now securely in place, I attached the covering for the slit, which has a 1.2 m × 0.6 m (4 ft × 2 ft) section at the bottom that flips out on hinges. The upper 3 m (10 ft) section slides over the top (see Figure 16.5).

I also considered other special needs when planning the "perfect" observatory which would soon house my 635 mm (25 in) f/5 Newtonian telescope. Local building regulations required that I build an engineered support pier to carry the $\frac{3}{4}$-tonne weight of the telescope. Apparently, there was no precedent for such a structure on the records of the inspection branch. Once I explained to the civil engineer and

Figure 16.5 The dome-slit, with lower flap and sliding upper section.

the building inspector just what I was trying to accomplish, I met with no resistance. I had our contractor put in a 1 m × 1 m (3 ft × 3 ft) pilon consisting of concrete blocks reinforced with steel rods and filled with concrete. It sits on a special footing anchored to bedrock and runs up through the three floors of our home into the dome (see Figures 16.6 and 16.7). To accommodate movement of the telescope inside the dome, I had our builder reduce the diameter of the pier to 0.3 m (14 in) where it entered the dome. The floors and ceilings in the rooms below butt up to, but do not touch, the pier itself, since any vibration caused by walking, doors closing, etc., would be echoed through to the telescope and camera. Also, since heatwaves cause distortion on the mirror, the telescope must be constantly kept at ambient temperature. I therefore arranged for insulation of the floor and the separate stairway leading to the dome. An interior door at the bottom of the stairs closes the office off from the stairs. I put in a lift-up trap door at the top of the steps, and having two independent doors between the dome and the warmer temperatures of the house has proven quite effective. (These days, I often turn on an exhaust fan to gently move air through the dome about an hour prior to doing any observing.)

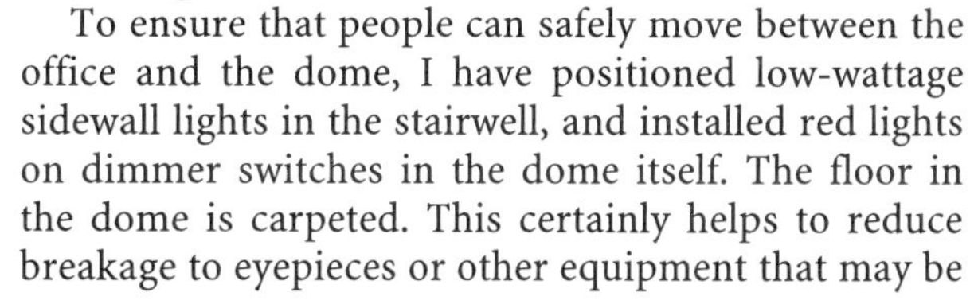

To ensure that people can safely move between the office and the dome, I have positioned low-wattage sidewall lights in the stairwell, and installed red lights on dimmer switches in the dome itself. The floor in the dome is carpeted. This certainly helps to reduce breakage to eyepieces or other equipment that may be

Figure 16.6 The position of the observatory on the building.

Figure 16.7 The concrete telescope pier (*just visible, left of centre*) rises through three floors of the house.

dropped from the top rung of my tall observing ladder. I have also affixed strips of the carpet to the inside of the dome wall to act as a protective skirting where the wheels run in the track. This not only prevents fingers, hands or clothing from becoming accidentally ensnared while the dome is being rotated, but offers the further advantage of cutting down the drafts which would normally enter through the gap. The dome sometimes does retain some moisture due to marine fog. I keep a small heater kept dialled to a very low setting in close proximity to the telescope and computers. During the winter months, high winds often buffet the mountain-top. To help prevent the dome from lifting, I have thick chains attaching it to the inside of this pony wall in at least a half-dozen places.

I built the telescope mount around a 430 mm (17 in) Mathis worm gear. The German equatorial design best suits my needs. The right ascension housing is constructed from heavy-walled 200 mm (8 in) pipe welded into 15 mm steel positioning plates. The right ascension shaft itself is 150 mm (6 in) pipe mounted through a bearing at each end; a larger clutch was added to this assembly incorporating a four-screw pressure plate. The declination housing is 150 mm (6 in) in diameter with a 100 mm (4 in) pipe shaft mounted in bearings at each end. The counterweight shaft is just an extension of the declination rod to secure the 450 lb (200 kg) of counterweight. The top end of the declination shaft is attached to a 0.3 m × 1 m (1 ft × 3 ft) 10 mm ($\frac{1}{2}$ in) steel plate. This plate has two larger rings welded to it to support the telescope tube. A 450 mm (18 in) tangent arm declination clutch is attached around the declination housing and worm-driven with a reversible synchronous motor. The cradle is 10 mm ($\frac{1}{4}$ in) × 50 mm (2 in) wide steel bands reinforced with 25 mm (1 in) angle-iron. The 55 lb (25 kg) base of the cradle is a 1 m × 0.3 m (40 in × 12 in) slab of 12 mm ($\frac{1}{2}$ in) steel. The mirror telescope tube and cradle weigh in at over 300 lb (140 kg). The finished mount is equipped with optical encoders on both right ascension and declination for a CAT (computer-assisted telescope) computer.

The 50 mm (2 in) thick Pyrex primary mirror for my 635 mm (25 in) f/5 telescope was produced by Galaxy Optics in Colorado, and has a beautiful figure. The guiding head and eyepiece provide me with nearly 800 power for guiding. The mirror cell is 645 mm (25$\frac{1}{2}$ in) pan-style, with a 10 mm ($\frac{1}{2}$ in) aluminium plate base and a 50 mm (2 in) band screwed around the base to support the sides of the mirror.

I floated the mirror 100% on bubble pack (plastic packing material with air bubbles). I secured the mirror with six claws, which do not touch the surface of the mirror. I had a shipyard fabricate the barrel for the 3 m (10 ft) long tube, which is 3 mm ($\frac{1}{8}$ in) thick aluminium, rolled in three sections and welded together. I have since painted that aluminium tube over with flat black paint, and find that the tube currents have been greatly reduced.

I use a 110 mm (4$\frac{1}{4}$ in) minor-axis diagonal and home-made off-axis guider. The simple guider is constructed using a 10 mm ($\frac{3}{8}$ in) aluminium plate that rotates in a ring mounting on the side of the telescope where the focuser would be positioned. This forms a photographic platform and permits quick changes in a variety of equipment. The guider has two prisms which are mounted on one side of the 60 mm (2$\frac{1}{2}$ in) focusing tube. The prisms will rotate through the field and the whole plate will rotate as well. This is coupled with a 3x barlow and a 12 mm eyepiece which produces 800 power for guiding. I can virtually pick up any star in the field to guide on. I have a 180 mm (7 in) Meade f/9 refractor mounted onto the side of the large scope. This refractor has excellent optics for CCD imaging. I also use a 305 mm (12 in) LX200 series Meade computer-controlled telescope.

I now use a number of CCD cameras, some of which I own and others which I am beta-testing. I use three computers for imaging and guiding my telescope. The first is a 486-66 with 32 mB of RAM and a 1.4 GB hard drive. The motherboard has a cache using 4 mB of RAM with the CD-ROM and *The Sky* program. I have an NEC 4GF monitor and Diamond Viper 24-bit graphics video card, utilising 2 MB on the card. The 486 computer is in my office, directly below the observatory and is connected through conduit to the observatory. This is designed to control my telescope from the warm room. I also have a 386-40 in the dome, which I use for operating my CCD imaging camera. A second, older 286 is used to control the autoguider CCD camera.

I guide the main telescope with a home built off-axis guider. This guider features a binocular set of prisms injected into the edge of the same field of view as the primary CCD camera uses. The prism assembly slides under my electric focuser. I am presently using an ST-6 to auto guide my telescope. The telescope's guiding head is designed to rotate 360° and slide in through the field at the same time. It makes picking out a star to guide on from the field a very simple task. Although I have a 180 mm (7 in) Meade refractor mounted on the side of my large telescope, flexure remains a problem. I have found it much easier to place a small 90 mm (3.5 in) Maksutov on the top of the Meade and use the ST-4 or ST-6 to autoguide the 17 mm (7 in) with the big telescope's drive. Even though this combination sounds very much like the "tail wagging the dog", it works flawlessly.

Using the *Sky Pro* program enables me to use three CCD cameras at the same time. I use one with *Sky Pro*, one with the 386 computer and the third one with the ST-6 to guide the CCD camera on the 286.

I'm having the time of my life in my home observatory. It's wonderful to be able to spend time observing, rather than facing a long drive to a distant site and a struggle to set up, only to have the wind or cloud move in. I can now image-process on the nights when the weather just doesn't co-operate, and love the new possibilities that CCD cameras and this wonderful observatory have opened up to me.

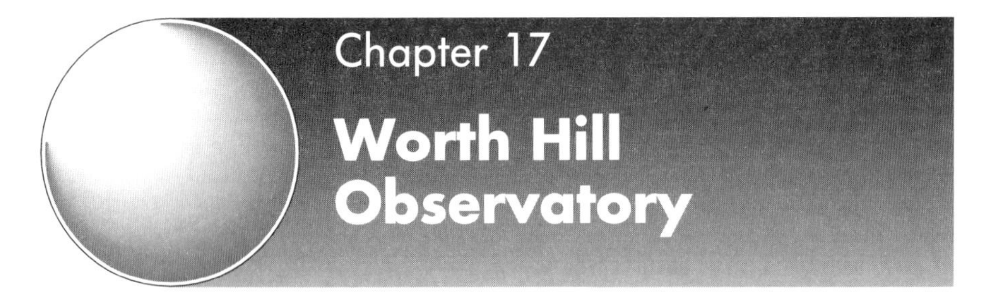

Chapter 17
Worth Hill Observatory

D. Strange

Introduction

Situated at the head of Seacombe Valley, Worth Matravers, Dorset, this observatory commands fine views over the English Channel, with an unobstructed sea horizon from east to west. This aspect also serves to protect my southern skies from the scourge of light pollution, allowing me fine views of deep-sky objects in the southern skies.

Figure 17.1 The open dome of the Worth Hill Observatory.

My interest in constructing a domed observatory was spawned by aperture fever – a desire to observe with larger telescopes with the aim of housing a permanently mounted 500 mm (19½ in) f/4 reflector. I live in an exposed site and needed a domed structure to shelter the telescope from strong winds. I decided to build an observatory 4.5 m (15 ft) in diameter with the idea of mounting the dome on a low concrete block wall 0.5 m (2 ft 6 in) high. This configuration allows me to observe objects of low declination in the southern skies.

Design and Construction

The first observatory I constructed was in fact a converted geodetic greenhouse, the frame of which I purchased as an ex-demonstration model from an agricultural show. I replaced the glass with Filon glass fibre sheets and although this worked well for ten years, the dome was eventually demolished by the infamous Great Storm of October 1987.

Having enjoyed the benefits of observing from a dome, I realised that I needed to replace, and with a stronger structure. I sought the advice of Derek Rolls, a fellow member of the Wessex Astronomical Society who happened to own a sheet-metal works.

We decided that the dome should be able to withstand storm force ten winds and would be built on the existing dwarf wall of the original observatory. We used a framework of 70 mm (2¾ in) semicircular angle-iron, which was mounted on a 100 mm (4 in) angle steel dome base-ring. The dome was constructed in two halves and clad in 40 segmented plastic-coated steel gores, which were joined together by clamping them with 25 mm (1 in) stainless steel strips. A rubber strip insert at each joint and bolts every 150 mm (6 in) keep the dome fully watertight. Although the resulting structure (see Figure 17.1) must nearly weigh 1000kg, it rotates easily on five heavy-duty all-steel wheels, and is kept on the dome track by five side-thrust wheels.

The 1 m (3 ft 3 in) wide dome shutter is made of three sections which run on eighteen Teflon wheels located in a 70 mm (2¾ in) channel either side of the dome slit. It is fairly easy to open and concertinas together when opened, allowing the telescope to view to the zenith and beyond.

Telescope and Equipment

The telescope (see Figure 17.2) is a 500 mm (19½ in) f/4 Newtonian mounted low down on a cast aluminium equatorial with a 400 mm (15¾ in), 720-toothed drive wheel on the RA axis. A tangent arm drive on the declination axis allows fine control of telescope elevation. Mounted alongside the main scope I have a 150 mm (6 in) Schmidt-Cassegrain which I use as a guide scope for photography.

For the past two years I have been using a Starlite Xpress CCD camera in conjunction with this telescope, which has given many exciting images of deep-sky objects. Luckily the dome has enough room to place a computer work station and desk at the northern end.

The CCD camera displays captured images on a CCTV monitor, allowing the computer to display the Hubble Guide Star Catalogue at the same time as imaging. This is a real boon to the art of "digital star-hopping" – enabling me to track down objects too faint to be seen visually. Using such techniques I have been able to image many faint quasars, Abell galaxy clusters, and comets including the faint comet train of comet Shoemaker-Levy 9 before it collided with Jupiter.

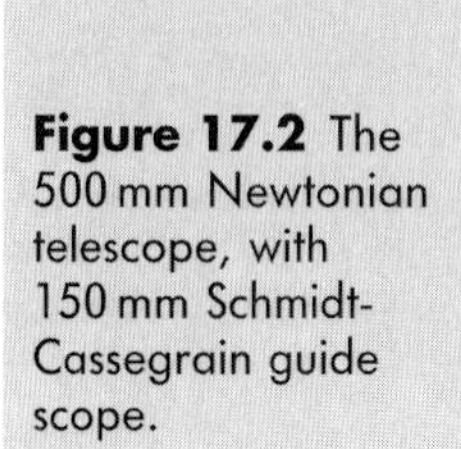

Figure 17.2 The 500 mm Newtonian telescope, with 150 mm Schmidt-Cassegrain guide scope.

Recently I have been doing various CCD photometry projects imaging variable stars with a Johnson's V filter. Computer communications adds a further interesting dimension to amateur astronomy and by such means I found myself involved in a project initiated by Dr. Kawai of Osaka-Kyoiku University, Japan, who put out an e-mail request for high time resolution photometry of the dwarf binary star SS433. The project called for photometric observations from observatories throughout the whole world in conjunction with X-ray observations from the ASCA satellite. It was encouraging for me to see that my results were of use, and due credit was subsequently given in the published paper.

A night's observing is now always preceded by a log on, either to Starbase One Bulletin board or Compuserve's Astroforum for an update on the latest astronomical discoveries and events. By this means images of comet Hale-Bopp were obtained within hours of the report by the discoverers themselves.

A fair proportion of observing time is also given to visiting school parties, cub and scout groups.

The observatory has now been operational for the past three years and is wearing well. Rather surprisingly the computer equipment has been very reliable, considering it is kept in a fairly harsh environment subject to varying levels of humidity and temperature and host to a wide variety of wildlife (earwigs regularly climb in and out of the keyboards!).

I am also pleased with how cool the inside of the observatory remains, even under soaring temperatures. I had imagined that an all steel structure would act as a heat sink and give rise to poor seeing conditions at nightfall. That does not seem to be so, since the white plastic-coated steel reflects heat well, and in fact some of the best planetary observing conditions have tended to occur soon after sunset.

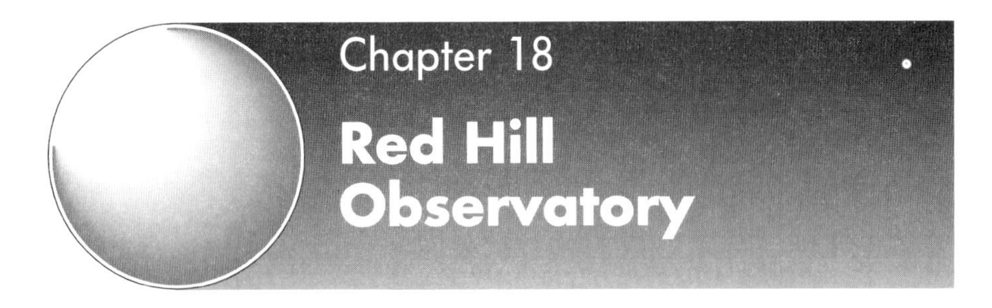

Chapter 18 Red Hill Observatory

Chris Plicht

My interest in astronomy started around 1965, when my brother and I started to ask questions about the stars. Around 1968 I bought a small Newtonian telescope with 76 mm (3 in) aperture on a shaky altazimuth mount. This instrument was carried around to suitable observing places and also was part of the luggage in the car for the family vacations in England and Italy. After being given a simple equatorial mounting by my father I was able to take pictures of the sky with short-focal-length lenses. But soon after 1971, astronomy no longer was my number one spare-time occupation - motorbikes and girls were much more fascinating.

Figure 18.1 Chris Plicht's observatory at Red Hill.

In November 1984 I bought a 200 mm (8 in) telescope of the Schmidt-Cassegrain type. This was in preparation for the return of Halley's Comet in 1986, which I planned to observe. Living then in a flat in Hildesheim, a town with about 100,000 inhabitants, I had to drive out of town for observations. This was quite inconvenient, because 30 kg (66 lb) of equipment had to be carried down the stairs into the car and back again after the observation at a remote site. In 1985 I rented a house in a small village. There I planned and built my first "shed" for the telescope. The main reason for this was that I really was tired of carrying all the equipment out of the house to a spot in the back garden, only to find clouds moving in immediately after I had aligned the telescope to the north celestial pole!

This first building was made of a wooden frame, planked with tongued-and-grooved boards. The roof was made of corrugated metal, moved away manually for observing. The telescope was permanently installed with its original fork mount on a concrete pier. The pier had a diameter of about 200 mm (8 in).

Halley's Comet was seen and photographed from this place on January 17th 1986. Later that year the telescope with field tripod and photographic equipment was taken to Tenerife, the largest of the Canary Islands. There, at an altitude of more than 2000 m (6000 ft) and under nearly perfect conditions, I observed Halley's Comet's brightest phase.

During a visit to Mount Palomar, California, on the occasion of our honeymoon in 1988, my wife said, "This looks like a real observatory – you should have a nice white dome, too."

So I got myself a commercially available dome (Baader Dome, 82291 Mammendorf, Germany) with a diameter of 2 m (6 ft) and a 400 mm (15 in) shutter. It was installed on the above mentioned wooden frame.

In 1989 I built my own house at an astronomically reasonable place and planned to erect the observatory again in the garden. But the observatory had a low priority, with other work needing to be done on the house and in the garden.

Eventually I began work. I planned to make the base building slightly bigger than the old one and ended up with a hexagonal design.

The first thing was digging out a hole for the concrete foundation of the telescope pier. It is about

1 m deep and 0.8 m in diameter. In May 1990 with the help of some friends I filled the hole with concrete and some reinforcing steel bars extending about 1 m above ground level. The pier was poured at a later date, using a moulding box made of plywood. It is 1.27 m (4 ft 2 in) high, 360 mm × 360 mm (1 ft 2 in × 1 ft 2 in) at the base and 260 mm × 260 mm (10 in × 10 in) at the top (see Figure 18.2). Three metal screws were set into the wet concrete. They would hold the wedge of the telescope.

The only way for me to build the dome base myself was using wood as a construction material. It is easily workable for the layman, it is available in almost any size and it is affordable. A real design phase did not exist: no calculations on wind forces or the weight of snow were applied to my ideas. A base hexagon was made by using eighteen boards each of 200 mm (8 in) width and 30 mm ($1\frac{1}{8}$ in) thickness. Six boards formed

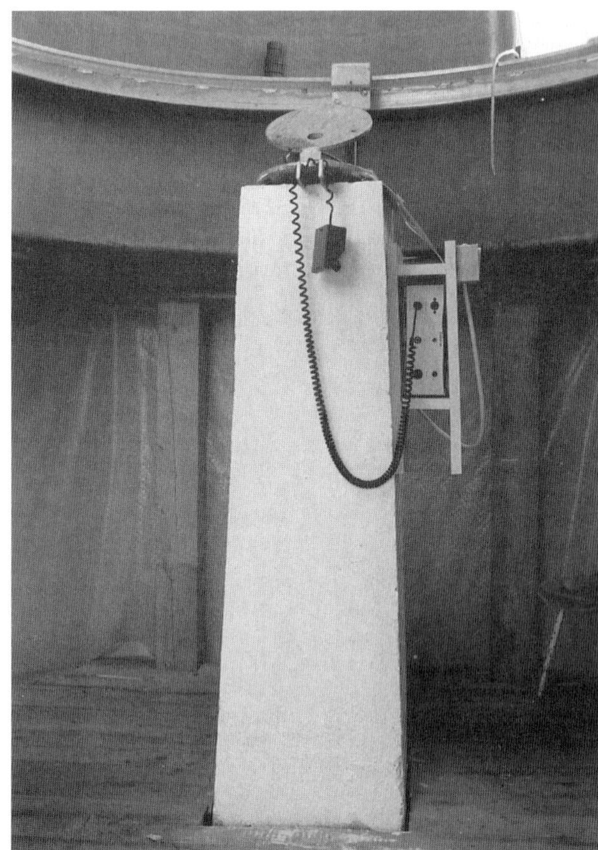

Figure 18.2 The concrete telescope pier.

one ring, and four rings were mounted one on another to give strength.

The top layer of the base hexagon had eight spacings for the beams that would carry the top hexagon and the dome. These spacings and the beams were 100 mm × 120 mm (4 in × 5 in) (see Figure 18.3). The base hexagon was put in place on a 100 mm (4 in) thick bed of gravel, protecting the wooden construction from water. Then the eight 1.1 m (3 ft 7 in) long beams were installed and the top hexagon was added. This was made from three layers of boards with the appropriate spacings in the lower layer to fit onto the beams. The sides of the base were first covered with thick reinforced-plastic foil and then planked with tongued-and-grooved boards. The outside of the base was then painted white to keep the temperature of the building low.

Two layers of tarred felt made the roof waterproof. The final work was done with the reinstallation of the dome; the parts had been awaiting their proper use for over six months.

The three parts of the cylindrical base were put on the hexagonal base and fixed with eight bolts on the top ring. This base as well as the dome is made of glass-fibre-reinforced plastic. The next step was the installation of an L-shaped aluminium ring. It rests on three rollers, allowing a full 360° rotation (see Figure 18.4). Three more rollers on the cylindrical base ensure that the dome stays centred while rotating. The

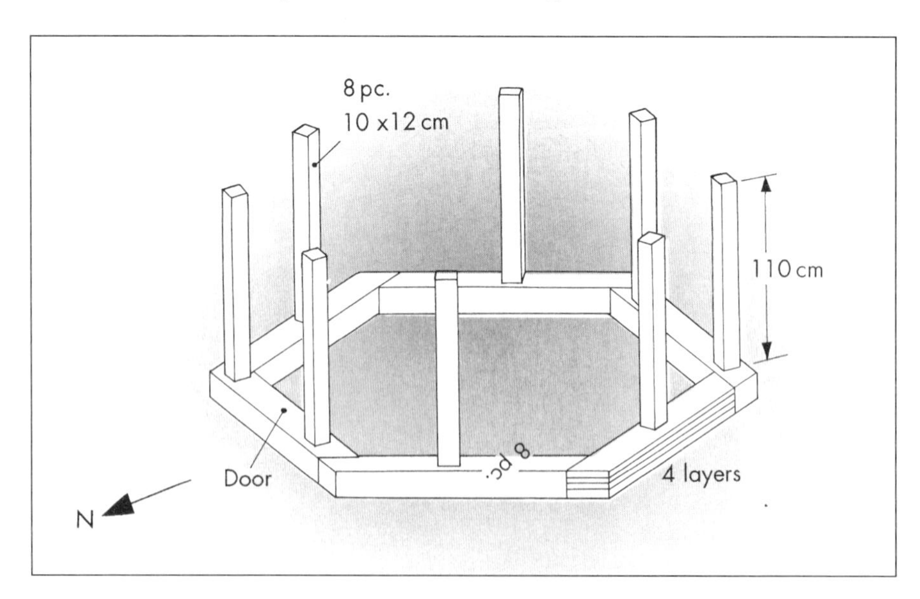

Figure 18.3 Plan of the observatory baseframe.

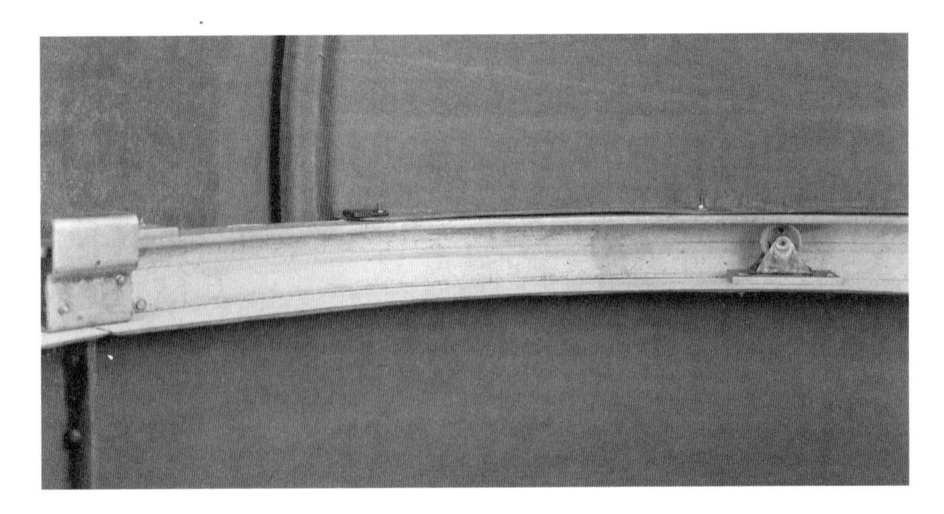

Figure 18.4 The L-shaped aluminium ring and one of the three rollers it rests on.

aluminium ring carries the two main parts of the dome. Adding the rear part of the dome, a sliding shutter and the front flap made the observatory complete (see Figure 18.1). All fasteners are made of stainless steel and all ropes are nylon.

Today, after nearly five years, some minor problems have become obvious.

First, the tarred felt-covered roof is not completely leakproof. Rainwater dripping through bolt holes and improperly sealed areas has caused the wooden floor to swell. I have had to reseal the roof and take special care of the places where the bolts hold the dome on the wooden construction. A complete new rebuilding of the roof with a slope seems to be too much work at the moment.

Second, the dome has a 30 mm ($1\frac{1}{4}$ in) wide gap around the cylindrical base. This allows for good ventilation, but as the observatory is sitting beside a wheatfield, dust also enters the dome. The telescope is covered in dust most of the time! To prevent the dust from entering the dome I am thinking about mounting a stiff cloth or rubber sheet to cover the gap.

Third, the interior of the dome is still in the same condition as it was right after construction. I am planning to have a permanent installation of red and white light sometime. The cable is already there, but not connected to the house circuit, so at present I have to roll out the extension cord whenever I go out for observations.

As I write this, in June 1995, some work on the outside of the building is due.

Weeds are taking over around the observatory, the outer boards need repainting, and a new lock has to be installed. However, I remember that when I started with the whole project I did not want to have a showroom, but rather an observatory meeting my needs. This goal was achieved.

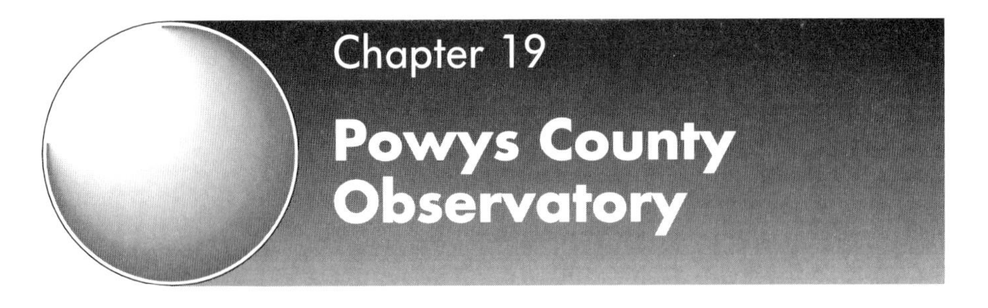

Chapter 19

Powys County Observatory

Cheryl Power

The Observatory enables anyone to experience first hand the fascination of observational astronomy. We hoped that inspiration and interest would be aroused in people and especially children who have had no real opportunity to sample the practical side of astronomy. Schools, universities and societies can take advantage of this unique facility, which will cover a wide range of astronomical subjects and include meteorology and seismology. These disciplines are all within the science curriculum. The observatory project is non-profitmaking. The site at Knighton is approximately two acres and is at an elevation of 417 m (1368 ft).

The Observatory building houses a 4.7 m (15 ft 6 in) planetarium to take up to thirty visitors. This

Figure 19.1 Powys County Observatory.

unit is an ideal teaching tool. It clarifies the mystery of celestial mechanics, and shows how to find your way around the night sky. It is a favourite with young and old alike. (We have even had requests from insomniacs to sleep in the planetarium!)

The Observatory has two domes. One houses the camera obscura, an instrument used for scanning and viewing the surrounding countryside and wildlife. This unit can also be used for celestial, solar and lunar projection. Cloud formations and sunsets are particularly magical to watch.

The large dome contains the main refractor, solar telescope and 165 mm ($6\frac{1}{2}$ in) refractor, and is designed to be used as an enclosed viewing system incorporating an optical window. This means that the occupants of the dome room as well as the delicate equipment can be kept in a warm environment. For serious work the window can be removed. The dome room floor moves to facilitate easy access to the prime focus.

The main telescope is a 340 mm (13 in) f/10 apochromatic triplet refractor, an ideal instrument for viewing the Moon and planets (see Figure 19.2). The elements are oil-spaced with crown glass at the front, a borate flint for the centre and an extra-dense flint for the back element.

Figure 19.2 The 340 mm and 165 mm refractors and (just visible) the solar telescope.

The solar scope images in the Hα part of the spectrum in a bandwidth of just 0.7Å. This telescope will enable anyone to see the activity of the surface of our star, thereby gaining an insight into far-off stars similar in makeup to our Sun.

A refractor exactly half the size of the main telescope and of a similar design is also housed in the main dome. The unit has one of the American, Roland Christian's, lenses.

The Observatory houses a number of smaller instruments along with computers and teaching equipment. A 250 mm (10 in) Meade catadioptric telescope is used as a portable unit, as well as a Celestron C5 and Meade 150 mm (6 in) Dobsonian-mounted Newtonian reflector.

A weather station has been installed to take in images from the Meteosat and GEOS geostationary and NOAA polar orbiting satellites. The Meteostat images come in every four minutes. Images of Europe can be sequenced and then animated so that the forthcoming weather can be predicted. The polar orbiting satellites transmit less frequently, but the images can be spectacular, especially during the summer months.

Because the observatory is sited close to a number of fault lines, including the Church Stretton Fault, a seismological observatory has been set up in co-operation with Liverpool University. This unit is now monitoring local disturbances as well as the earthquakes throughout the world.

Although the observatory was built to be used primarily as a teaching unit, we hope that anyone interested will use the facilities. We want to ensure that everyone has an exciting as well as informative visit. Much will be done to achieve a mix of awe, excitement and interest, and a desire to learn more. We want all our visitors to leave the observatory enlightened, inspired and looking forward to another visit.

The observatory was built by Brian Williams and Cheryl Power to their specific design. The building took over four years for them to complete. They also made the optics for the main refractor and the camera obscura which took a further year.

On the 6th July 1995 Dr Patrick Moore kindly officially opened the observatory. The many sponsors who had donated their goods and services to the project were delighted to meet him. Members of the recently formed Offa's Dyke Astronomical Society were also at the opening and presented Patrick Moore with a specially engraved tumbler.

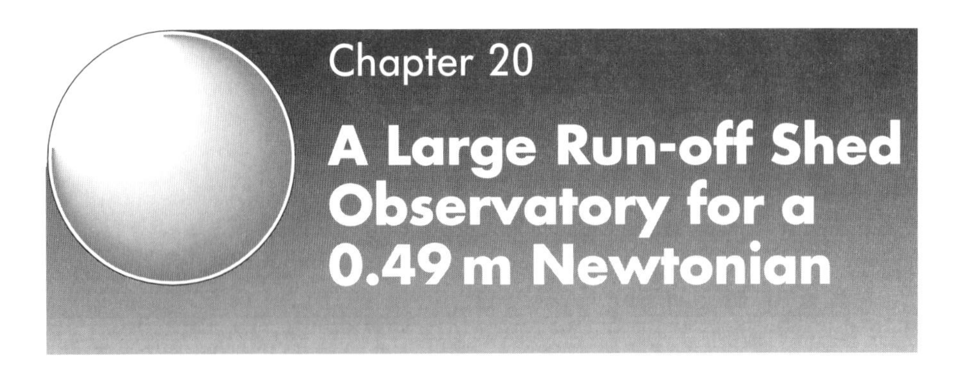

Chapter 20
A Large Run-off Shed Observatory for a 0.49 m Newtonian

Martin Mobberley

Introduction

Since 1980 I have worked as an electronics engineer in Chelmsford, England. Chelmsford is some forty miles to the south of the village of Cockfield in Suffolk, where my original 0.36 m (14 in) f/5 – f/20 Newtonian-Cassegrain telescope resides.

Figure 20.1 The shed rolling back to the west.

This telescope is in the garden of my parents' home, an excellent dark site well away from any street lights, but forty miles away from me!

Obviously, having a telescope sited some forty miles from where you live is not an optimum situation. However, as I lived in a first-floor flat until 1991, using the 0.36 m at Cockfield at weekends was the only way I could do any serious observing.

In 1991 finances allowed me to move into a bungalow on the outskirts of Chelmsford and I immediately set about siting my second telescope. Finances were not my only consideration in planning my move for 1991. The CCD camera revolution was just dawning at that time, and for the first time I could anticipate getting good results from a light-polluted town site.

Telescope Specification

As soon as I began house hunting in early 1991 I gave my required telescope specifications to Astronomical Equipment Ltd of Harpenden, and my mirror requirements to their optics manufacturer, AE Optics of Cambridge. These companies were run by two brothers, Rob (equipment) and Jim (optics) Hysom, who also built my 0.36 m reflector in 1980.

During the 1970s, 80s and 90s the Hysoms supplied more 0.25 m to 0.6 m (10 in to 24 in) Newtonian telescopes to British universities and polytechnics than anyone else. This was to be Rob's last large telescope before he retired from telescope making, and I was thus particularly proud to own it; I know that Rob was pleased that his last big 'scope would go to someone who would use it. He was able to use spare parts from other projects in completing the instrument, which brought the cost down considerably.

After a number of discussions we decided on a 0.49 m (19 in) f/4.5 Newtonian on a massive German equatorial mounting with 76 mm (3 in) steel RA and declination shafts, and a 0.45 m (18 in) diameter phosphor bronze worm wheel for the RA drive.

As well as the main telescope, I acquired a number of other British-made components. These included a 125 mm (5 in) refractor guide telescope, also supplied by Astronomical Equipment, quartz-locked sidereal drive electronics by Astrotech, and large (300 mm; 12 in) diameter aluminium setting circles and a super-smooth helical focusser from Astrosystems.

I considered a number of possible instruments for this second observatory, but two requirements determined the final choice, namely:

1. I wanted the largest aperture I could afford.
2. I wanted the best sidereal drive I could afford.

The outcome was the half-ton monster shown in the Figures.

The 0.49 m telescope was ordered in March 1991 and delivered in March 1992.

Observatory Design

Due to the large size of the instrument and my dislike of domes, I decided that the observatory should be of the run-off shed variety. Although domes are the traditional observatory buildings they do have a number of disadvantages, namely:

1. They are difficult and expensive to build or purchase.
2. They prevent the observer from being "under the stars" and seeing bright meteors etc. or cloud rolling in.
3. Long exposures necessitate moving the dome slit with the telescope.
4. Domes take a long time to cool down after a hot day, causing atmospheric turbulence.

The main advantage of a dome is the protection from the wind – especially bitterly cold, energy-sapping wind!

I learnt a number of lessons after building the run-off shed for my 0.36 m (14 in) reflector, most of them related to the difficulty of doing things in the dark when it is cold and damp. I wanted the shed for the 0.49 m (19 in) reflector to be a substantial structure more than able to withstand gale force winds from any direction. It had to be totally waterproof and prevent as much wind-blown dirt and rain getting in between the shed-rail gap as possible.

Despite the size and implied weight of the shed it had to be very easy to roll to and fro on its rails – this factor had been a major problem with the 0.36 m run-off shed.

The problem with any shed that has to have one side "removed" when it rolls back is potential loss of structural rigidity, i.e. the shed may flex and bend as it moves. In addition, if the wheel/rail interface is less than perfect the shed can grind to a halt and even jump off the rails.

Having had personal experience of trying to prise one corner of a "jack-knifed" half-ton shed out of thick mud at 3 a.m. in winter, I can assure you that it is not recommended!

So, the wheel-rail design is highly important. I have now tried three different types of wheel and rail over many years and the final solution, as employed in the 0.49 m shed is by far the best.

Rail Design

My first wheel-rail design was naive beyond belief: simple nylon wheels on the base of the shed and flimsy L-section angle-iron on which the wheels ran. The rails rested on the grass between posts which held the angle-iron in position! After several disastrous nights with this shed, late in 1980 the L-section angle-iron rails were screwed firmly to 100 mm (4 in) cross-section wooden beams fixed to the posts. Further refinements included setting the posts in concrete and reinforcing the shed to prevent excess flexure.

The final refinement was to totally dispense with the L-section rails and employ U-section, i.e. the wheel track was totally enclosed rather than just restricted on one side.

This solution is generally reliable, although if a stone gets stuck between the wheel and the U-section it can cause the shed to jam. Brushing the U-section rails of this shed was one of my regular chores.

The problems I experienced led to a complete rethink for my second shed. The most reliable solution was really quite obvious: use a deep V-groove pulley wheel and an inverted T section rail (see Figure 20.2). Such rails are easily obtained in long lengths from scrap metal yards and are easily screwed down

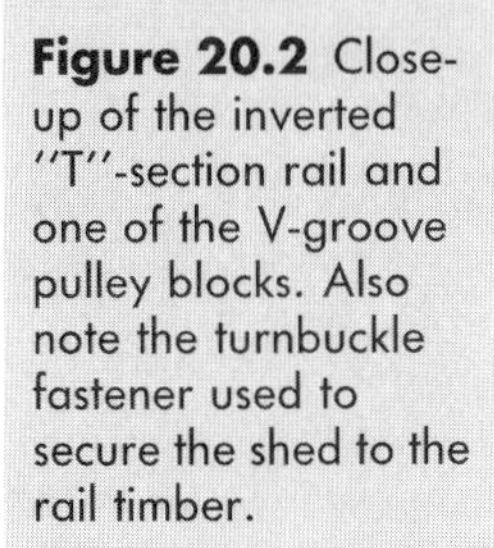

Figure 20.2 Close-up of the inverted "T"-section rail and one of the V-groove pulley blocks. Also note the turnbuckle fastener used to secure the shed to the rail timber.

to timber supports. A V-groove pulley wheel restricts the shed's sideways motion and the shed can never jump off the rail. Pulley wheel units can be obtained from many hardware stores.

An additional advantage of this design is that leaves and stones cannot cause any problems. The only aspect of this design which needed care was ensuring that the inter-rail distance was precisely right and precisely the same along the length of the rails. This was easily achieved by delaying the final "screwing-down" of the rails until the completed shed was running freely.

Shed Design

My 0.49 m (19 in) f/4.5 Newtonian has dimensions as follows:

- Tube length (including mirror cell) 2.3 m ($7\frac{1}{2}$ ft).
- Telescope width from outer edge of tube, through tube, to end of tube counterweights 1.8m (6 ft).
- Telescope height (base of equatorial mount to top of tube, with tube horizontal) 1.1 m ($3\frac{1}{2}$ ft).

Obviously, the shed needed to be longer, wider and taller than the above dimensions. A major consideration in the "shed size equation" was the height of the concrete plinth on which the telescope sits.

I purchased an ideal "ready-made" plinth from a nearby builders' merchants, in the form of a concrete drainage pipe 0.56 m (22 in) in diameter and 1 m (3 ft) deep. When filled with concrete, such a plinth will weigh over 1.5 tonnes; a sizeable plinth for a sizeable telescope!

As my 0.49 m (19 in) Newtonian is on a German equatorial mounting, an overly large plinth diameter could prevent the telescope reaching the zenith; i.e. the mirror end of the tube could clip the plinth for objects between 30° and 70° declination. Luckily, the drainage pipe is not quite large enough to cause this problem. Another plinth consideration which affects the shed height is, of course, the proportion of the plinth above the ground. The overall height of my 0.49 m reflector, packed away in the shed with the tube horizontal is 1.1 m plus the height of the plinth and baseplate (see Figure 20.3).

The 125 mm (5 in) refractor guide telescope attached to the 0.49 m (19 in) Newtonian does not affect the height consideration; it is slung beneath the main telescope when the observatory is closed.

A high plinth does enable a large German equatorial telescope to track further past the meridian (in either direction) without the telescope or guide telescope hitting the ground.

Conversely, a high plinth makes it more difficult to reach the eyepiece without long ladders, when obser-

Figure 20.3 Eastern doors opened to show the 0.49 m (19 in) f/4.5 Newtonian in the stowed position.

ving objects near the zenith. A compromise plinth height (above the ground) of 0.45 m (19 in) was settled on, giving a worst-case eyepiece height of 2.7 m (9 ft), i.e., reachable with a standard aluminium stepladder.

The baseplate (which interfaces telescope to plinth) was supplied to my drawings, by Rob Hysom of AE. With the plinth decision made, work started on the observatory in October 1991.

The first step was to dig the 0.55 m deep × 0.56 m (about 22 in) diameter hole for the concrete drainage pipe. Steel rods were hammered deep into the hole base to provide additional rigidity for the concrete which would fill the pipe. With the help of six colleagues from work and two substantial lengths of timber, the drainage pipe (weighing 150kg (330 lb) unfilled) was manhandled onto the back lawn and, carefully, slid into position.

A couple of days of mixing and pouring concrete into the pipe then followed.

When the concrete had reached 15 cm from the top of the pipe, concreting was suspended until the arrival of the baseplate.

In September 1991 I had sent detailed drawings of my run-off shed requirements to a local shed manufacturer in Chelmsford; they agreed to build the custom shed for a reasonable price. The shed was to have an internal width (across the eastern opening) of 2.6 m ($8\frac{1}{2}$ ft), so that it would be 0.3 m (1 ft) wider than the tube length. It would be 2.3 m ($7\frac{1}{2}$ ft) deep, so that it would be 0.5 m (18 in) deeper than the expected tube to counterweight distance. The eastern door would be 1.6 m (5 ft 3 in) in height to allow the shed to clear the telescope tube (the rails plus wheels add a further 100 mm (4 in) of safe clearance in this respect. The shed roof could afford to slope down to only 1.4 m (4 ft 7 in) above ground level at its western end as this end only needs to clear the counterweights and not the main telescope (see Figure 20.1).

A double, hinged door was to be provided for the eastern side; the heavy two-stage removable side had proved to be a big hassle in my first shed design. Hook fasteners on the north and south walls were provided for securing the large hinged doors when the observatory was open. The shed, in kit form, was delivered to the observatory site in October 1991, just as the plinth was being completed.

Choice of Site

In choosing a house – six months earlier – for siting a large telescope, the suitability of the garden for a large telescope was of paramount importance. The house satisfies a number of vital requirements:

1. The garden faces due south and the southern aspect has no obstructions above about 18°.
2. The house, some 4 m (13 ft) to the north of the run-off shed, is a bungalow and so does not restrict the northern aspect above 20°.
3. The house is sited to the south of Chelmsford such that the worst light pollution is to the north. (There are more interesting targets in the southern half of the sky.)

The land to the south of the house slopes downward starting from the middle of my lawn so, to get the maximum height over southern trees, the shed was mounted close to the house.

One advantage of this is that computer/CCD equipment can easily be wheeled out of the house on a trolley, rather than sitting permanently in a potentially damp environment.

As the prevailing wind, in the UK, blows from the west, it was decided that the shed should roll back to the west so that it could afford some protection for the telescope.

This strategy has certainly paid off on many breezy nights, although it is only completely effective when the telescope is looking at low-altitude objects and therefore fully shielded.

Completion

Following the delivery of the shed to the observatory site in October 1991, the shed assembly was completed within a matter of days. Three V-groove pulley blocks were attached to the base of the north and south walls of the shed (making six in total) as a first step. Two 6.5 m (21 ft) lengths of timber (treated with creosote) and T-section rail were already prepared in advance of the shed delivery to enable the shed to be assembled while on the rails. Considerable care was taken to ensure that the rail timbers were perfectly horizontal along their length and with respect to each other; this was achieved by using wooden shims under the rail timbers.

Once the shed was painted, screwed together and a felt roof added, the final inter-rail spacing was determined and angle-iron corner posts were hammered into the lawn and screwed to the rail timbers. The T-section rails were then screwed firmly to the timber.

The shed is prevented from rolling about on the rails by turnbuckles anchored to the inside of the timber rail supports and attached to hooks on the inside of the shed base (see Figure 20.2). A total of four turnbuckles is employed, two on each rail. Dust and dirt entering the shed (between rail and shed wall base) is minimised by a plastic skirt all around the base of the shed.

The large double door features two bolts (one into the roof and one into the ground) as well as a large padlock. An extremely loud and piercing burglar alarm is fitted to the door and has to be disabled within seconds of the door opening.

The baseplate, designed to interface plinth to telescope, was collected in December 1991 and found to fit perfectly inside the lip of the pipe. The final 150 mm (6 in) of concrete were then poured into the plinth top and the long rag-bolts provided with the baseplate were pushed deep into the setting concrete.

The final task was to lay a plastic conduit a foot below the lawn for the electrical power supply cable, fed from the house garage. Four mains sockets were screwed to the plinth using rawplugs to supply power for the telescope and CCD equipment.

Finally, the whole area under the shed was levelled and paving slabs were added.

Delivery and Conclusions

The telescope was delivered to the site on March 26th 1992. Due to meticulous planning, liaison with the telescope manufacturer, and my experience with the first shed, installation went smoothly and without a hitch. First light was on the night of the delivery of the telescope, when I was treated to a superb view of Jupiter and its satellites under almost perfect seeing

conditions. There were no unforeseen problems and, given the chance to rebuild the observatory, I would do it the same way again.

Figure 20.4 The telescope plus author ready for observing; the shed is at the end of its westward travel. Note the CCD equipment trolley, which has been wheeled out from the house.

Chapter 21
Tenagra Observatory

Michael Schwartz

Inception

I had three motivations for building Tenagra observatory:

1. I was building a house in the country and building the observatory simultaneously would obviously make things easier.
2. My romance with astronomy was changing. I was no longer interested in hauling equipment and imitating a bad contortionist at the eyepiece. The advent of CCD cameras added more wires and some hair-trigger booby traps. It was time to permanently mount the scopes.

Figure 21.1 Michael Schwartz's observatory at Tenagra.

3. I needed to permanently house a new generation mounting (with very precise pointing capabilities) so a long-time interest, extragalactic supernova hunting, could finally be realized.

The supernova hunting system meant using a telescope that automatically slewed and imaged many galaxies per night. Putting the scopes in a dome meant that automated movement of the dome would be necessary. I have other reasons for not wanting a dome, but a design that avoided this problem was a definite asset. Simple is best. So I chose a sliding roof design.

Telescopes Housed

Tenagra observatory has an ArchImage mount that is fully automated. It cannot be moved by hand. The mount is completely controlled via computer in the control room, as are the CCDs and video cameras. It carries two optical tube assemblies simultaneously. In my case it holds a custom 317.5 mm ($12\frac{1}{2}$ in), f/3.5 Newtonian and an 280 mm (11 in) f/10 Schmidt-Cassegrain. The Newtonian has a "permanently" attached CCD camera. The Schmidt-Cassegrain takes a variety of accessories, from visual to CCD to video, as well as a 102 mm (4 in) piggyback refractor (see Figure 21.2).

The Observatory Design and Concept

I decided that Tenagra Observatory would have a control and a scope area. The choice then was how much room to allocate to each. Since some construction costs are fixed and one-time, I decided that the control room concept would be extended to "play room", so I designed the control room as large as my budget could support.

So in addition to virtually all telescope controls, the 5.8 m $\times$ 4.6 m (19 ft $\times$ 15 ft) control room (see Figures 21.3 and 21.4) houses a small music studio and archery equipment, including an area for arrow making. It also has, of course, a place to sleep. The

Figure 21.2 The ArchImage mount holding the $12\frac{1}{2}$ in Newtonian and 11 in Schmidt-Cassegrain.

Figure 21.3 The control room and (*in background*) the scope area.

control room is electrically heated for winter use and air-conditioned for summer.

An Aside: Building for All Contingencies

I kept in mind that there may come a time when I would no longer live in this house and that the observatory and control room would be part of an eventual sale of the house and grounds. Therefore all construction is up to local building codes and the style, roofing material and siding match those of the main house. Since I had to dig trenches for telephone lines and power I also placed water lines for an eventual bathroom.

It makes no sense to open a trench and not include all possible "feeds". The control room could easily be a large study or even studio apartment. And while I tiled the observatory floor because it would be continually soaked by dew, it has been suggested that with the pier removed the scope room would make a great hot-tub room with a roll-off roof! Of course it was a greater expense to build the structure according to all local building codes, but the alternative in the event of a sale of the house and property meant that the observatory is a usable (and saleable) asset rather than a sub-standard liability.

Design of Tenagra Observatory

Figure 21.4 shows the floor plan of the scope and control rooms. The main entrance is marked and is most often used, given the control room's multiple purposes. There are two entry doors to the 3.5 m × 3.4 m (11 ft 6 in × 11 ft 2 in) telescope room. The first door is for access from the control room and the second leads directly to the scope room. The wall between the control and scope rooms has a large window that allows me to comfortably watch the scope from the computer consoles. While I never had any real problems with cables (CCD cameras, dew heaters, video cables, etc.) I never feel comfortable when a high-torque mount slews around without me checking to see that it doesn't get itself into trouble. I have visions of snapping cables and control boxes dragging around in circles! So the window allows a more than full view of the scopes and mounts in all positions.

The building itself is exactly oriented to the compass points and the ridge peak of the control room running east-west with as low a ridge profile as possible (see Figure 21.1). My latitude is just shy of 45° north, so an angle of about 22° for the roof gives me a full view of all circumpolar stars at different times of the year.

I don't consider the obstruction a problem for two reasons. First, I have similar angular obstructions (mountain ridge and trees) to the east and west. Second, my average seeing is far from exceptional and I often prefer to observe only above 30° altitude whenever I can get away with it.

The next choice was the most difficult. I have a relatively unobstructed view to the south (including the ecliptic) and wanted to take advantage of this. If I had opted for the traditional horizontal roll-off roof then I would have limited this view. On the other hand, construction of a horizontal roll-off is relatively simple. I opted for the view, and the associated construction problems as can be seen in Figure 21.1. The

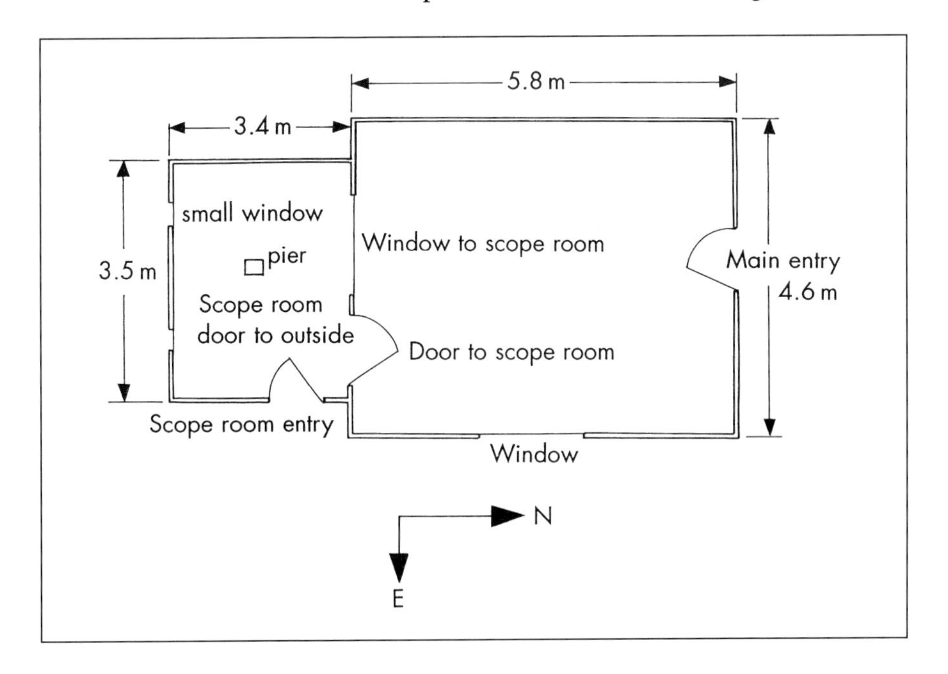

Figure 21.4 Floor plan (not to scale). All dimensions are internal.

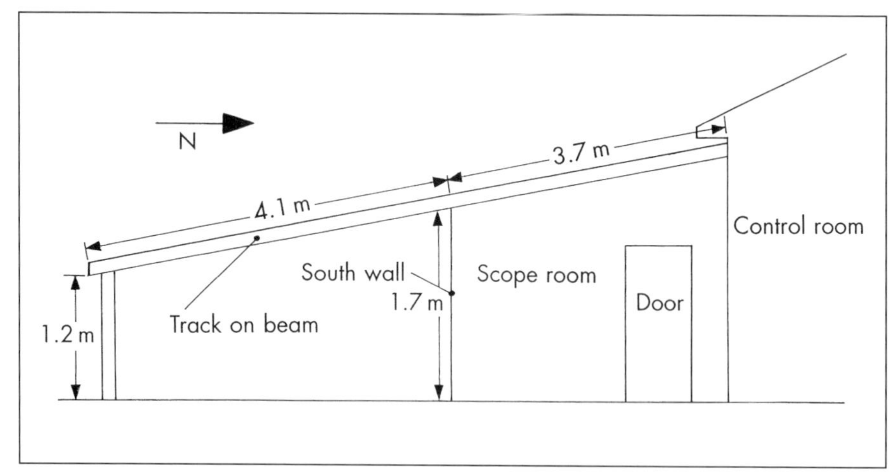

Figure 21.5 East section of roll-off (not to scale).

angle of the roll-off is about 10°. Figure 21.5 shows the overall design and dimensions.

There were two main problems that needed to be addressed once I decided to slope the roll-off roof. The first is careful design to get just the right angle to maximize the telescopes' views (i.e. less clearance between scopes and roof when the scopes are parked), properly dovetail the roof with the main structure of the control room, and carefully measure angles for the structure that would hold the roof when it is rolled back. While this may seem trivial, I even went to the extreme of building a wooden mockup of the mount (which had not arrived) and the OTAs. This was placed on the cement portion of the pier to verify the calculations I made to determine the optimum angle.

The second problem with an angled roll-off is roof weight. The third is its secure anchoring when closed. Weight is the main problem. I didn't want a moving roof that would require counterweights or a complicated locking and unlocking procedure for opening or closing. I have wet air in the summers and Pacific Northwest winter rains. So the materials would have to be completely waterproof. I decided to throw away all traditional standard wood construction techniques and use the simple design shown in Figure 21.6. The frame for the roof is made of aluminum 102 mm × 51 mm × 3 mm thick (2 in × 4 in × 1/8 in) rectangular tubes. This was welded by a not-so-local shop and moved to the building site (with great nervousness) tied to the side of a rental truck.

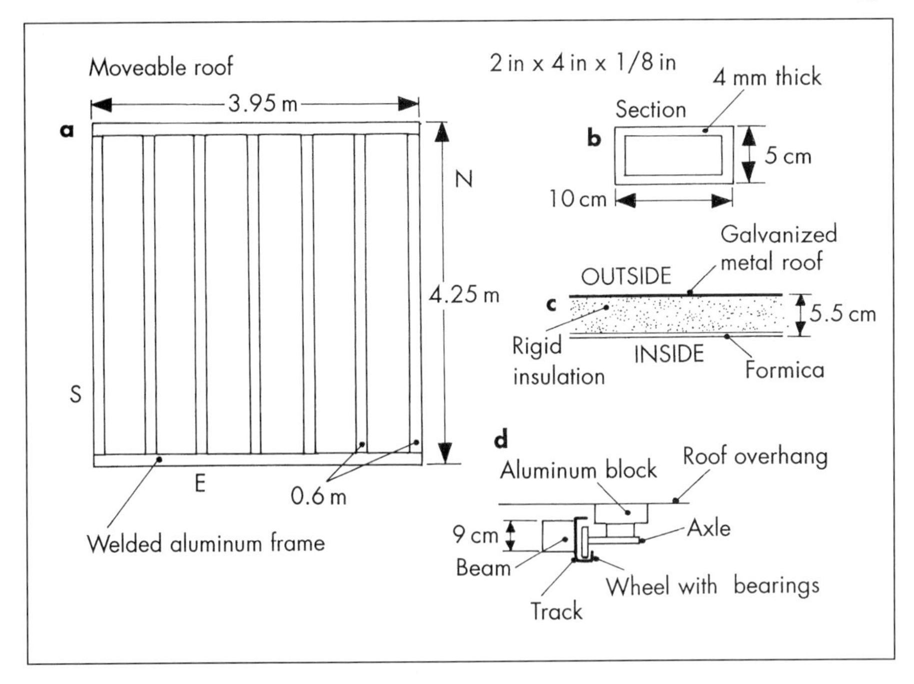

Figure 21.6
a Plan of roof frame.
b Section of aluminum.
c Section showing roof composition.
d Wheel and track assembly.

Once the roof frame was on site it was placed on sawhorses, and rigid insulation was placed in all openings. Then the frame was tapped for screws and flexible Formica was screwed down for the inside covering. A galvanized metal roof was screwed down on the other side. See Section c in Figure 21.6. The objective was accomplished: the roof was easily lifted by four people and placed on top of the telescope room. This design produced the desired light weight, fully insulated and weatherproof moveable roof.

Given the light roof, it was now time to decide how to move it and where to obtain the parts. In keeping with the spirit of over-engineering, industrial garage door straight tracks with matching ball-bearing wheels were used. This method locks the wheels in the tracks (there is virtually no up and down wheel play in the tracks) and serves as the necessary anchor in high winds. See Section d in Figure 21.6. The system is virtually frictionless. Figure 21.7 shows the tracks with the roof down and the scope peeking above. There are seven wheels on each side of the roof.

Once the roof was on the tracks I found that I could, with a little effort, move the roof up and off the telescope room manually and unassisted. Although this is not something I would like to do at

Figure 21.7 The tracks with the roof down.

5:00 a.m. with any regularity, it was nice to know that I could do it in a pinch.

I opted to use a standard winch with a hand controller for raising and lowering the roof (see Figure 21.8a). This small winch, rated to lift 450 kg (1000 lb) vertically, is much more powerful than is needed to move this lightweight roof at its slight angle.

I did discover, though, an interesting effect when the winch was connected by cable to the roof. Many people don't realize that the roof of a building is responsible for much of the structure's integrity. If the roof is not fixed you have wobbly walls.

When I initially hooked up the winch and stopped it before it made contact with the bumpers at the end of the track, the wall on which the winch was mounted would wobble in a frightening manner. This was handily taken care of by placing a U-bolt where the winch cable connected with the roof and placing springs (see Figure 21.8b) to allow for movement in the roof rather than in the south wall. If I have to stop the roof prematurely the halt of the inertia is absorbed by the springs rather than the wall.

The winch is solidly attached to the south wall using connected aluminum plates on the inner and outer sides of the wall. Finally, note in Figure 21.8b that there is considerable overhang to protect the winch (and the rest of the building) from the elements.

a

b

Figure 21.8 a The winch used for raising and lowering the roof. **b** The U-bolt cable-connection and shock-absorbing springs.

Internal Design Considerations

As with most observatories, the scope and control rooms were literally designed around the permanent pier. My pier is 1 m × 1 m (3 ft 3 in × 3 ft 3 in) and 1.3 m (4 ft) deep. It is heavily reinforced with steel, including the portion that is above the scope room floor (see Figure 21.9).

Note that the pier is a combination of concrete on the bottom and a 460 mm (18 in) steel pier. Some

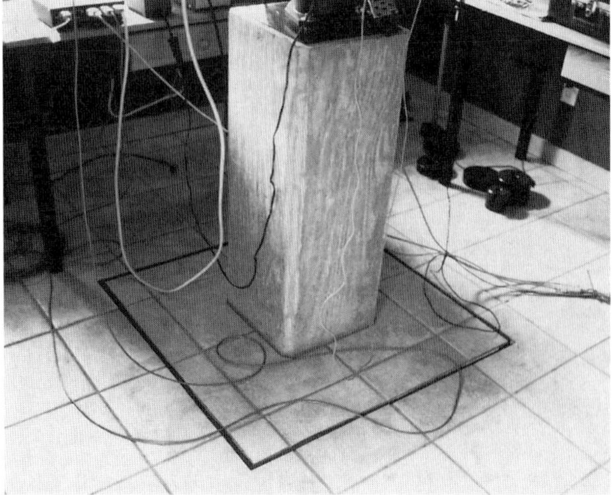

Figure 21.9 The telescope pier.

people would consider this too massive, but I wanted to absolutely eliminate vibration even if I added 140 kg (300 lb) of additional equipment. Again, there are one-time costs (e.g. making concrete forms) and there is very little additional cost for a few additional cubic meters of concrete.

Additional planning of the telescope room was done keeping two things in mind: (1) to use materials that assume the room has no roof and everything is exposed to the elements; (2) to develop a simple system that creates a constant mild positive pressure to keep dust and pollen from collecting on exposed optics and other surfaces.

I painted the interior walls with an external oil-based house paint and covered the concrete floor with exterior grade tiles (see Figure 21.9). While the latter certainly is not necessary, I had to consider that an average Pacific Northwest night can go through the dew point several times and everything open to the air is drenched. A wooden floor, or a floor covered with industrial carpet, would only hold water to the point where getting to the telescopes would be like walking through a swamp. I also have two large fans that run continuously when the roof is closed. Not only do they carry cool or warm air from the control room's heater and air conditioner, but they create a slight positive pressure in the scope room that tends to keep the room dust- and pollen-free as well as moderating the temperature. The fans also dry any dew on the floor and scopes after a night's observing. It beats mopping up. The two small windows on the south

side of the scope room (see Figure 21.1) allow substantial outflow of air.

Post Mortem

What would I do differently? Very little. I have thought that it would have been nice to build the observatory in an area with less obstructions. Originally I had figured that I wouldn't want to image anything less than 30° altitude. Since then I've gotten nice images towards the south at as low as 8°. But a different location would mean that I would have to get into the car and drive off to another site where it would have cost enormous amounts to get power etc. And once I had an observatory that is a stroll from the house I got spoiled.

There are still nights when I'd like to get out the chainsaw and fell the 200 trees that block my view in the east, but they are just too nice to sacrifice! So I wait an extra hour to take a look at M33. Moving the mountain a little to the west is equally unthinkable. So I'll have to wait until spring to search for supernovae in Leo.

Although I first thought that the window between the scope and control rooms was a necessity, it has turned out to be something of a problem due to the open-truss design of the Newtonian. When in just the right (wrong) position, the primary or secondary can catch a red light from the computer, spoiling a CCD exposure. This was easily eliminated by either keeping the full moon shroud on the scope at all times, or by placing a lightproof shade over the window.

There are a couple of features that could be problems if my observatory was not rural. Incident light, as discussed above, would be a problem. In addition, my current use of the cable and winch is rather noisy and could upset a neighbor. Virtually all of the noise is due to the gearing in the winch, and since any winch will suffice for this design it is probably possible to find one that puts out fewer decibels.

Tenagra observatory was a labor of love and a demanding construction experience. But I intended that the structure would last as long as the new house. A temporary observatory is one that breaks into pieces in order to move it to another site - a permanent one should last longer than the observer.

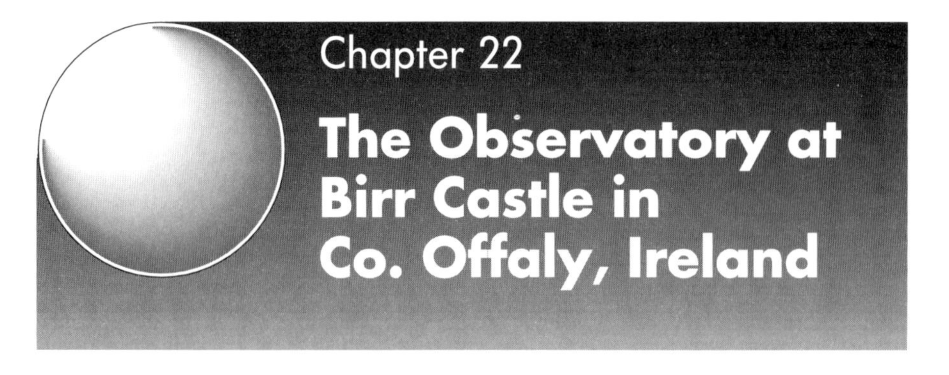

Chapter 22
The Observatory at Birr Castle in Co. Offaly, Ireland

Patrick Moore

It is safe to say that the observatory built and used by the third Earl of Rosse, at Birr Castle in County Offaly, Central Ireland, is unlike any other constructed either before or since. It was unique, but it was responsible for a major advance in astronomical science.

At that time the largest telescope ever built had been made by William Herschel; it had a 49 in (1.2 m) mirror and a focal length of 40 ft (12.15 m), and it had been fully steerable, though admittedly it was clumsy

Figure 22.1 The stone-wall observatory and telescope tube at Birr Castle.

to use (most of Herschel's work was carried out with telescopes of much smaller aperture).

In the 1830s Lord Rosse, an Irish landowner, made a 36 in (0.9 m) telescope on the same pattern, and it worked well. Lord Rosse then decided to make a really large telescope – a reflector with a mirror no less than 72 in (1.82 m) in diameter.

Lord Rosse was not a professional scientist, though he had graduated from Trinity College, Dublin, and was well versed in optics. The telescope was made at Birr Castle itself; Rosse had no help apart from workers on his estate, whom he trained. The metal mirror was cast (this even involved building a forge), but the mounting and the "observatory" had to be carefully considered.

In view of the limitations of engineering techniques at the time, it was evident that an attempt to make the 72 in fully manoeuvrable would have ended in failure. Wisely, Lord Rosse recognized this. The "observatory" consisted only of two massive stone walls. The tube of the great telescope was pivoted at the lower end, and therefore could swing for only a very limited distance to either side of the central meridian; the observer had to wait for the Earth's rotation to bring the target object into view! The optical system was Newtonian, so that the observer was placed on a platform high above the ground; it looked unsafe, but in fact there were no accidents throughout the whole of the telescope's first period of activity, from 1845 until 1909. Obviously it needed a team to control the movement; it was never possible to install an efficient drive, and there were no efforts to use the telescope for photography.

Despite its cumbersome pattern and its limited scope, the telescope was a triumphant success.

With it, Lord Rosse discovered the spiral nature of the objects we now know to be galaxies, and made many other valuable observations; it is said that nobody ever went to him for help or advice and went away disappointed, and there were, at various times, many eminent observers at the Castle.

After the death of the third Earl, his work was carried on by his son, the fourth Earl, but by the end of the century the telescope had been superseded by instruments of "modern" type, and it is true that the most fruitful period of its career ended in the 1870s. It was dismantled in 1909 after the death of the fourth Earl, and the mirror removed to London.

However, the stone-wall observatory remained *in situ*, together with the tube.

Finally, from 1996, the telescope was restored and brought back into use – admittedly as an historical exhibit only; no attempt was made to use it as a research telescope. However, it has its honoured place in history, and the unique Birr Observatory stands as a monument to the energy and skill of its maker.

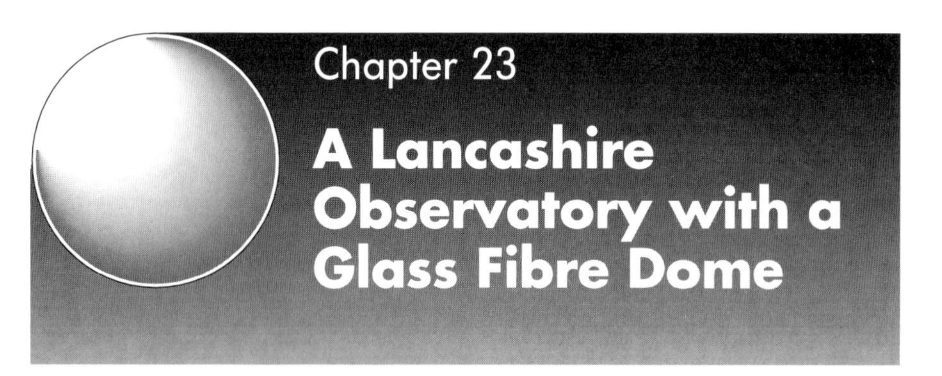

Chapter 23
A Lancashire Observatory with a Glass Fibre Dome

David Ratledge

Introduction

I had built many telescopes over the years, starting with a 150 mm (6 in) Newtonian and working my way up to a 320 mm ($12\frac{1}{2}$ in) Newtonian by the late 1970s. The drawback with my early ones was having to carry the telescope outside and set it up before observing could begin. Equally discouraging was dismantling it at the end of a long night. My interest then was in astrophotography (it is CCDs now) and this dictated accurate polar alignment and a massive mount to carry all the astrophotography equipment that inevitably becomes attached. A permanent observatory was the obvious answer.

The requirements for my observatory were onerous. It would have to withstand the damp (very damp!) Lancashire weather. A homogeneous roof, with no joints to let the weather in, was essential and would ideally be made from a material that required no maintenance. In addition, I was surrounded by street lights so it would also have to mask them as much as possible. Protection from the wind – not just for my comfort, but to stop any shaking of the telescope during a photograph – was also important. It would be in my garden so it would need to look attractive too – I didn't want an eyesore. All of this pointed to a dome in either aluminium or fibreglass.

Figure 23.1 David Ratledge's glass fibre observatory.

I looked at aluminium silo tops, but at around 5 m (16 ft) in diameter they were too big for my needs. However, I had become familiar with glass fibre construction when making telescope tubes, mirror cells and other components for the various telescopes I had made. I had been buying the raw materials from a boat-making company and there didn't seem much difference between making a glass fibre dome and a boat. Both had to be pretty good at keeping the water out!

Compared to the complex shape of a boat, that of a hemispherical dome is easy. So a fibreglass dome it would be.

Design

The first design stage of my observatory was to work out what diameter and height it needed to be. Although it was to take the 320 mm ($12\frac{1}{2}$ in) f/6 Newtonian I had just finished, I had made the fork mount large enough for a 450 mm (18 in) telescope. Therefore the observatory had to be capable of allowing for that eventually. Scale drawings (see Figure 23.2) quickly identified that I needed walls 1.5 m (5 ft) high with a diameter of 2.75 m (9 ft) making an overall height of nearly 3 m (10 ft).

Incidentally, the fork mount (see Figure 23.3) is ideal for a dome in that it minimises the required size. It is more compact than the German type of mount which carries the telescope tube to one side. A dome would also provide full headroom throughout.

The biggest problem I had to solve was that of how to make the circular track. It is traditional to use a circular steel rail on a loadbearing wall on which the dome rotates. There are two problems with this. First, there is the circular steel rail itself. It was beyond my capabilities to make one and it would have to be bought in. Second, it would require a loadbearing wall, which I wanted to avoid.

Golfball rotation systems (a ring of golfballs carrying the dome) overcome the rail problem, but not the loadbearing wall requirement. The reason for me wanting to avoid a loadbearing wall was partly cost, but mainly the fact that a solid wall builds up heat during the day. This heat is dissipated slowly during the night, causing image-degrading air currents. I wanted to use glass fibre for the walls so there would be no heat buildup and it would not need extensive foundations.

The novel solution I hit on was to use a fibreglass track, mounted on the dome rather than on the walls. The track would rotate and the wheels would be fixed in position on top of four posts. The timber posts would carry the weight of the dome and so the walls would not need to be particularly strong and could therefore be made of glass fibre.

Construction

The technique of making components with glass fibre is relatively simple. A mould has first to be made onto which glass fibre matting is placed and then thoroughly impregnated with resin. When set, a shaped-glass-reinforced plastic moulding of high strength and low weight is produced. The required thickness is built up by applying the necessary number of

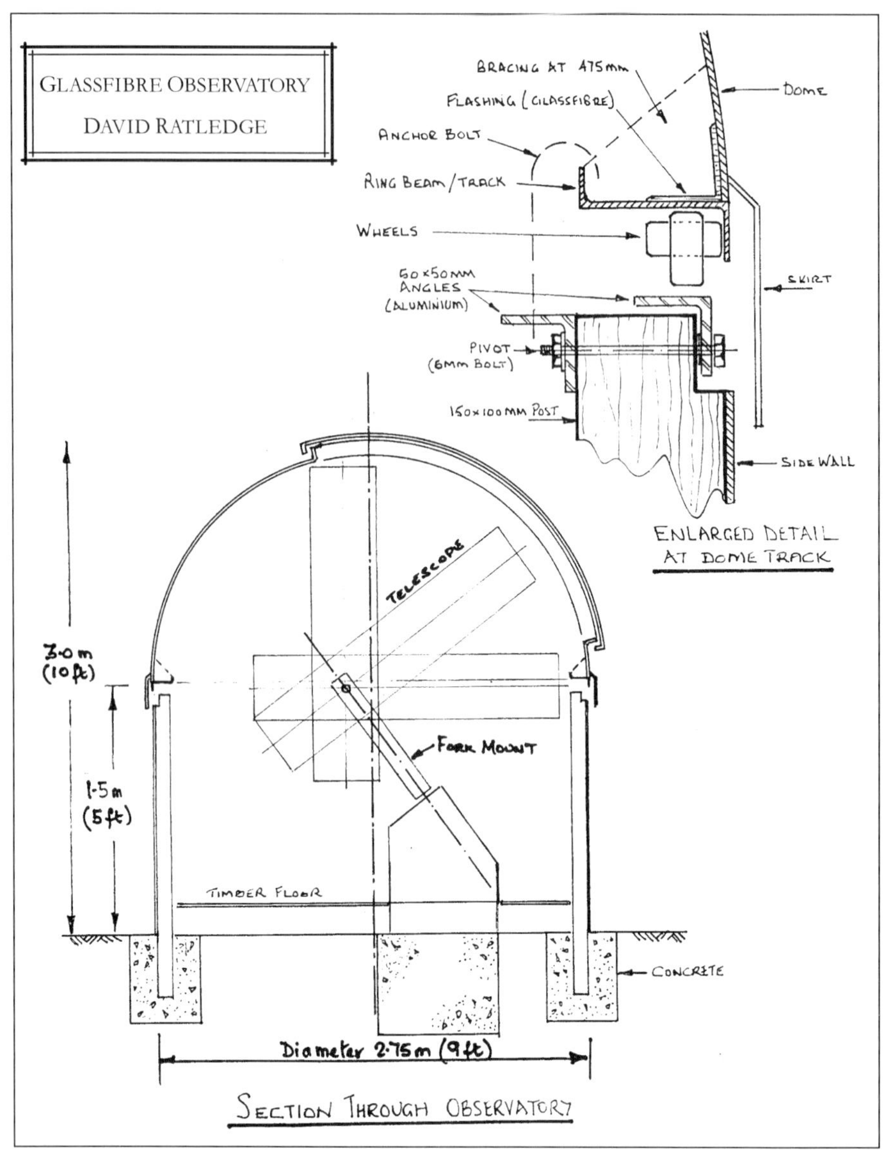

Figure 23.2 (*This page and opposite*) Scale drawings for the glass fibre observatory.

layers. It is easily sawn or drilled and is both weatherproof and rot-proof.

To impregnate the matting, a special roller with steel washers (discs) is used. This enables large pieces to be made quickly. Smaller intricate pieces are impregnated using ordinary paint brushes. Bought in bulk from boatyards, glass fibre matting and resin is reasonably cheap. For boat-building a mould is usually hired and later returned after making the components. However, I knew of no one hiring out

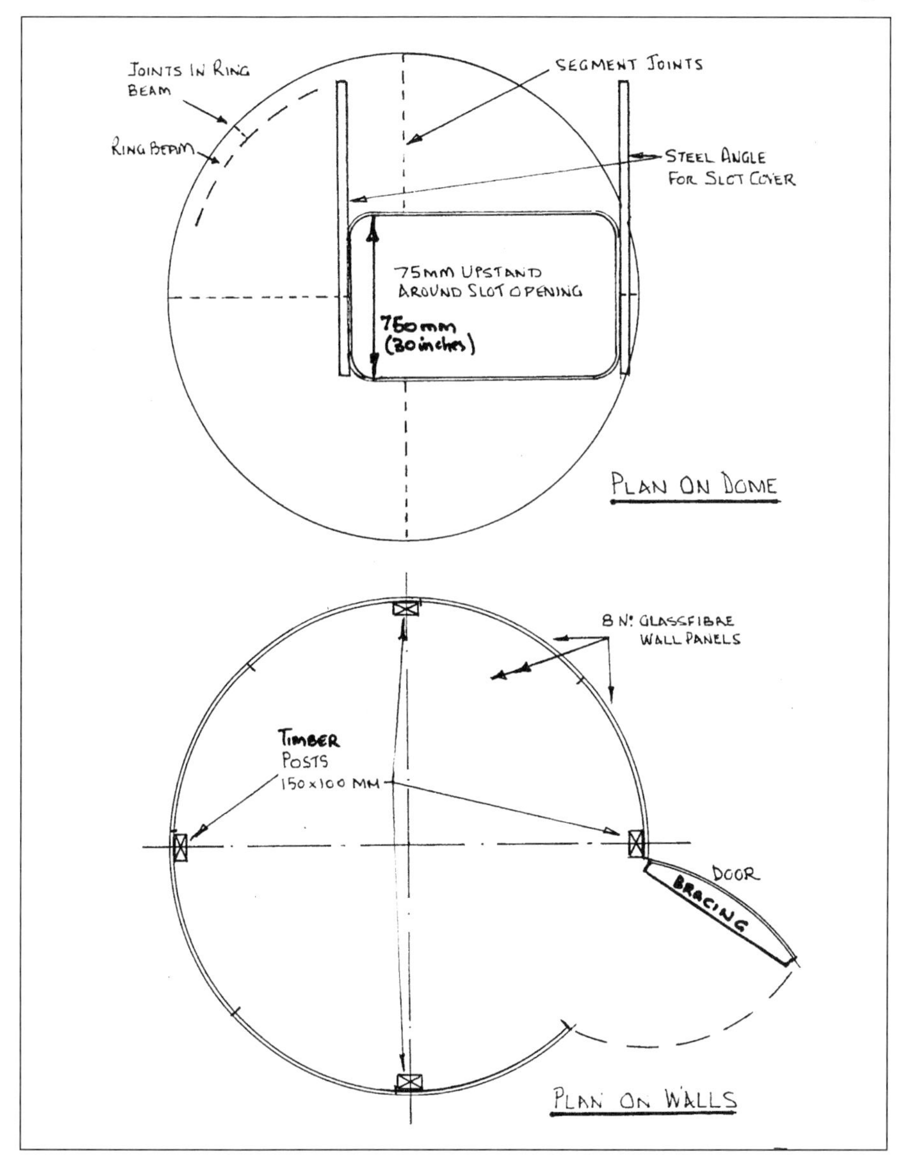

moulds for observatories – so that would have to be made too.

A mould for a hemispherical dome is curved in two directions and cannot therefore be made from flat sheets unless a faceted approximation is acceptable. I wanted a true hemisphere and so a doubly curved mould was required. However to build a mould for the full dome is not practical, except for very small ones or where a large production run is envisaged. The practical solution is to build a mould for just a

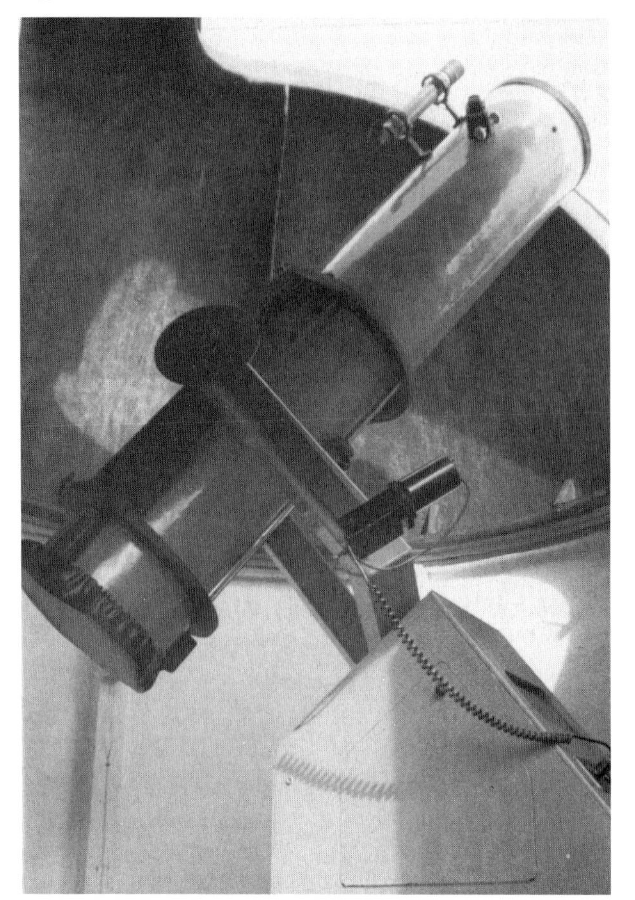

Figure 23.3 The 320 mm Newtonian on its fork mount.

segment from which the appropriate number of panels would be struck. But what segment – a half, a quarter, an eighth? I realised that a quarter had some unique advantages. It has three identical sides and three right-angle corners (if you think this is impossible from your school geometry days, remember that a dome is a three-dimensional object). It would therefore be the easiest to make, with some guarantee that it would all fit together.

The mould was made with plywood formers covered with chicken wire (wire mesh) and finished with builders' plaster trowelled smoothly to the curvature. It was sealed and coated with mould release. The four quadrants were made with a small lip on two edges. Without the lips the quadrants would be too floppy to handle, but care is needed not to trap the casting on the mould, making release difficult. I didn't at this stage worry about an opening.

Figure 23.4 The butt-jointed ring-beam segments (with block of wood in centre for checking circularity).

The next job was the ring-beam. Again, quarter segments were made. This was a Z (or double L) in profile (see Figure 23.2). This shape is structurally better than an angle or flat. It acts as the track and it provides surfaces for both the vertical wheels which take the load and the horizontal wheels which locate the dome sideways. It was made of double-thick glass fibre matting. The mould was made with plywood and hardboard. The four ring segments were then butt-jointed together in my garage and packed so they were level and circular. A piece of wood was glued to the floor in the centre of the circle and a piece of string attached to it and used to check circularity (see Figure 23.4). When I was happy, the joints were tacked together with small pieces of fibreglass just to hold them in place while the dome segments were positioned on top. A small hole was cut with a jigsaw in two of the dome segments where the final slot opening would be. This was to enable me to get in and out when joining everything together. The joints in the dome segments were staggered relative to the ring-beam segments. This is important for structural integrity.

The dome segments were temporarily joined with self-tapping screws through the small lips. This held them while all the joints were flashed over, inside and out, with strips of fibreglass both between the dome segments themselves and then between the dome and ring-beam. Braces (see Figure 23.2) were added around the ring-beam. These were simply plywood covered in glass fibre. When fully hardened the 750 mm (30 in) slot was cut with a jigsaw. It was cut with radiused corners – about 150 mm (6 in) radius – as square ones are much weaker, creating areas of high stress and therefore a risk of cracking. The edges around this opening were reinforced with an upstand, which was made partially on the ring-beam mould and the rest of it against hardboard forms. The upstand strengthens the dome and also keeps out rain water. Finally, a layer of fibreglass and white coloured resin was placed over the assembled unit, thus completing the structure and making a homogeneous unit (see Figure 23.5). The whole structure is immensely strong and can easily carry my weight in the centre.

The slot cover is a single-sideways sliding unit, again in fibreglass. It is, however, singly curved and was easily made on a mould using a 3 mm ($\frac{1}{8}$ in) plywood sheet curved to the correct radius over plywood ribs. Lips were cast all around. They were 100 mm (4 in) on three sides, but only 25 mm (1 in) on the long side that had to slide over the opening. Two steel angles (rescued from an old bed) were used for the track, with grooved pulley wheels mounted on the cover. The cover is light and does not disturb the balance of the dome when it is opened (see Figure 23.1). When closed, it is secured by three rubber rings

Figure 23.5 The completed dome, with upstand round the slot and fibreglass/resin finish.

and hooks. It has sufficient overlap on the opening side to prevent rain blowing in.

The next job was the walls. These are non load-bearing and their fibreglass panels were made in a similar way to that described for the slot cover. In this case eight segments were made. One was braced on the inside and forms the door.

The dome weight is taken by four posts, at quarter points, on which the wheel assemblies are mounted. I used 150 mm × 100 mm (6 in × 4 in) timber for the posts. They were soaked in creosote preservative for several days before being concreted into the ground (see Figure 23.6). Each wheel assembly comprises three wheels, two vertical and one horizontal and is mounted on an aluminium angle which is pivoted on the posts (see Figure 23.7). This equalises the load on each wheel.

It took six people to carry the dome and place it on these supports. It looked very odd sitting on four posts, as the walls has not been added at this time, but it rotated smoothly, so the wall panels were added straight away. The reason for doing this quickly is that domed observatories actually generate downforce in a wind so that when the walls are in place, they become remarkably galeproof. The problem is that if the wind gets inside them then they can take off! So until the walls were in place, I worried about every gust of wind. For the same reason, the slot cover must be securely anchored to the dome when closed, for if it comes off in a gale then the structure is vulnerable.

Figure 23.6 The four posts on which the dome rests, each topped with a wheel assembly.

Figure 23.7 The wheel assembly: three wheels mounted on a pivoted aluminium angle.

All that was left to do was to add a skirt around the dome covering the gap between the dome and walls (see Figure 23.1). Again this was made with strips of glass fibre cast on the slot-cover mould. The skirt was finished in a contrasting colour, which added the finishing touch.

Conclusions

The observatory was completed in 1981 and is still functioning perfectly today. It has never leaked and has been virtually maintenance-free. It just needs the occasional wash.

One fortuitous aspect of the design only came to light later. When observing during freezing damp weather, ice forms on the inside of the dome. As this melts the next day, water runs down the inside but is caught by the ring-beam upstand and is trapped there. It eventually evaporates without doing any internal damage.

I have since added a false floor inside the observatory. This provides better access to the telescope eyepiece and, being made of plywood, is considerably warmer underfoot.

One thing I did not consider originally was a 240-volt mains power supply. With the advent of CCDs and computers (even lap-tops need mains power in the cold), I should have run a permanent power cable underground into the observatory.

The only other change has been one of colour. The observatory was originally white, but this made it very prominent and, with crime rising, I decided to camouflage it and changed the colour to green. It blends in with the background better and now no one knows it is there.

With a permanent base, the minute the clouds clear – and they do sometimes in Lancashire – then I'm ready to go, with the minimum of hassle. Over the years the observatory has proved very successful and allowed me to pursue my interest in astrophotography in probably one of the worst locations for it in Britain! Many of my astrophotographs have been published in the USA and UK, both in popular magazines and in books. With the advent of CCDs requiring a computer at the telescope, and with worsening light pollution, the case for a classical type dome is, I believe, stronger than ever and fully justifies the effort involved.

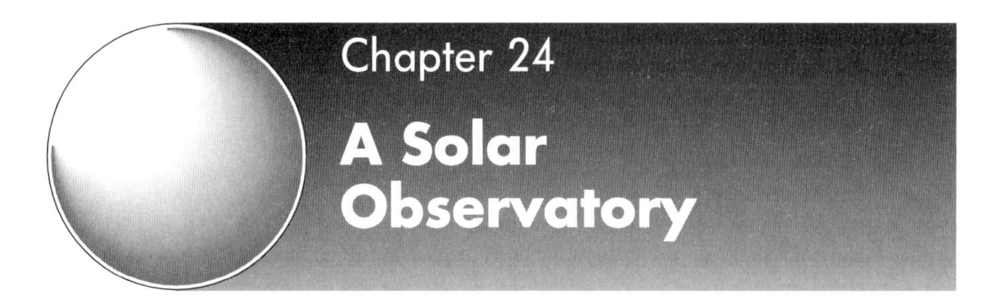

Chapter 24

A Solar Observatory

Eric H. Strach

Introduction

When observing the projected image of the Sun it is essential to exclude extraneous light in order to increase the contrast. When I first started regular solar

Figure 24.1 E.H. Strach's solar observatory.

work in 1969, I used a projection box attached to the eyepiece of a 75 mm (3 in) refractor. I obtained good results in favourable weather conditions but accurate positional work was impossible when the box was shaken by strong winds.

Having seen the description of W.H. Baxter's solar observatory in his book *The Sun and the Amateur Astronomer*, I decided to build a darkened observatory as the advantages seem overwhelming: it provides complete protection against weather, it offers wide access to the projection screen instead of the small aperture of the projection box and it also provides a permanent housing for the instrument and accessories, making it unnecessary to carry the telescope to the observing site, the fixed equatorial mounting remaining undisturbed.

I built the observatory in my spare time in 1970 and I have been using it ever since. Over the years I added some improvements but the basic design has not been altered and has served me well over the twenty-five years.

Design

Choosing the site was not difficult; the northernmost part of my garden provided a good outlook to the south and east, though regrettably a less satisfactory view of the west. In mid-winter my house hides the Sun from 1 p.m. onwards, restricting observing possibilities to 3 hours from 10 a.m. to 1 p.m.

Choosing the size was more difficult. I settled on a diameter of 2.75 m (9 ft) for the base-ring structure. The foundations were dug to a depth of 750 mm (2 ft 6 in) and filled with broken bricks, stones and concrete.

The observatory is essentially an octagonal building with a rotating conical eight-sided roof. (See Figure 24.1.)

Seven wooden frames were assembled to form a ring wall, the eighth section being the door. Each frame carried a window. A horizontal plywood ring was constructed and fastened to the top of the window frames and the door frame.

Eight weight-bearing wheel assemblies were fitted to the lower plywood ring: they carry the upper plywood ring and with it the rotating roof. Four centra-

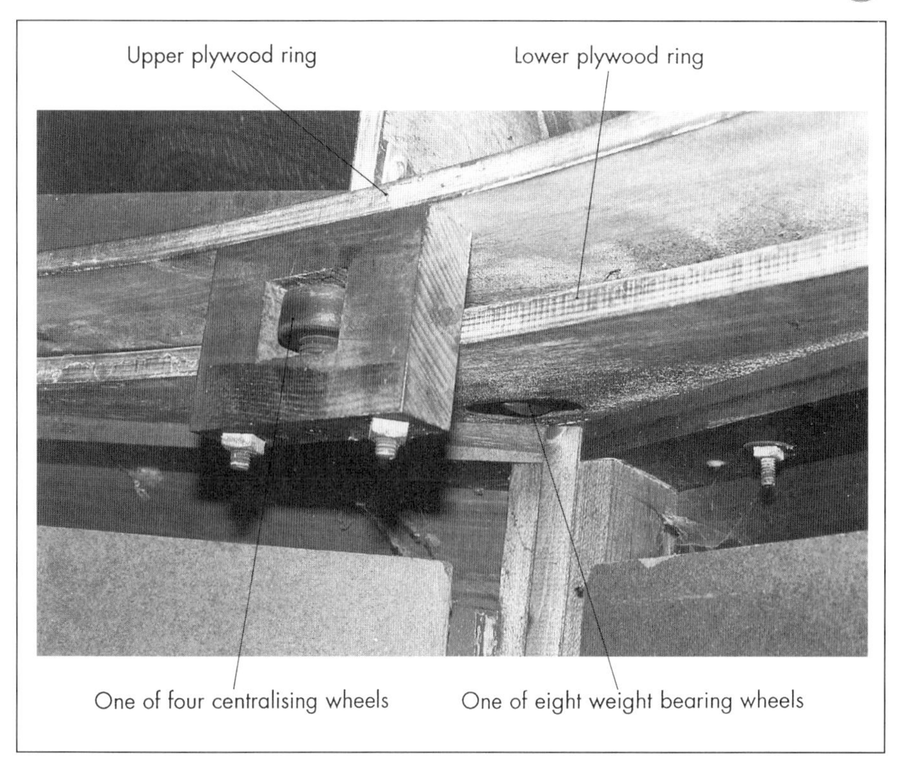

Figure 24.2 The weight-bearing and centralising wheel assemblies.

lised assemblies keep the roof on track when rotating, as illustrated in Figure 24.2.

The base of the roof consists of an upper plywood ring; it carries the conical roof sections which consist of marine ply attached to a framework. The roof contains a double shutter.

The observatory was designed to contain a massive wooden tripod with a German equatorial mount, carrying a 77 mm (3 in) refractor to which a projection screen is permanently attached. A central pillar carrying the equatorial head would have been a preferable alternative to the tripod.

The main material used for the building was wood, hence the expense involved was very reasonable.

Construction

The massive concrete foundation was essential to provide a firm base, free of vibration. A smooth surface was fashioned and plastic sheeting was incorporated to provide a damp-proof course.

Each of the seven wooden frames measured 1.07 m × 1.07 m (3 ft 6 in × 3 ft 6 in), and they were constructed in my workshop. The door frame measured 1.72 m × 760 mm (5 ft 8 in × 2 ft 6 in). All were erected to form the base-ring and firmly fastened to the concrete floor from which they were insulated by a damp course.

The seven wooden frames were filled in with cedar wood strips to give it a pleasing appearance and make it durable.

Seven robust window frames 940 mm × 600 mm (3 ft 1 in × 2 ft) were fixed to the tops of the side sections. A wooden door completed the whole ring structure.

The rotating roof supporting system consisted of two horizontal rings of 15 mm plywood ($\frac{5}{8}$ in), the lower ring being fastened to the top of the window frames and the door frame. Eight roller skate wheels made ideal means of supporting the upper plywood ring and allowing it to rotate. They were fastened to the upper surface of the lower plywood ring and allowed to protrude into its recesses (see Figure 24.3). The upper ring carries the roof and is separated from the lower ring by 32 mm ($1\frac{1}{4}$ in).

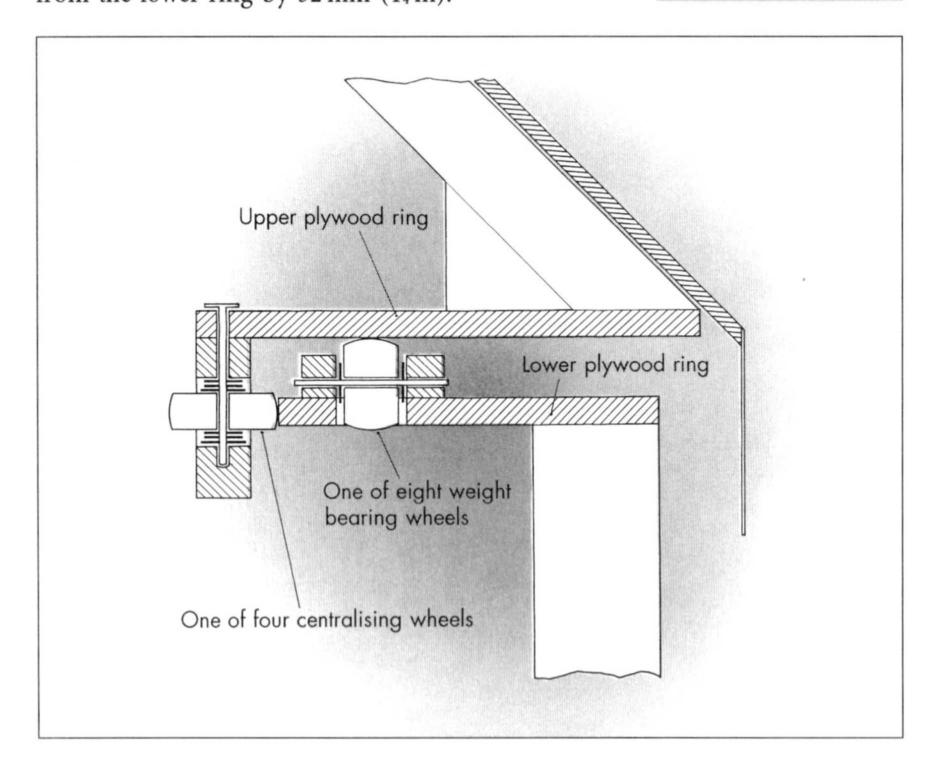

Figure 24.3 Cross-sectional plan of the weightbearing and centralising wheel assemblies

The rotating conical roof was more difficult to construct and was cut to shape from marine (resin-bonded) plywood in a joiner's shop by a bandsaw. One section of the roof contained the frame for a double shutter. They were mounted on rising hinges to open outwards; they were kept open by stays and were secured when shut. Two roller blinds were fitted to the shutter aperture, one at the bottom, the other at the top. They were connected by a light wooden frame, 0.6 m (2 ft) square, providing the only access of the sun to the interior. It could be positioned as the need arose.

In order to keep the roof section on track when rotating, four centralising assemblies attached to the upper ring protrude down from its inner surface. It is essentially a block of hardwood, 32 mm ($1\frac{1}{4}$ in) thick, 90 mm ($3\frac{1}{2}$ in) long and 130 mm (5 in) wide. It contains a horizontal wheel engaging the inner rim of the lower ring, as illustrated in Figure 24.4.

It took me three months to construct and assemble the observatory. The glazed window frames were covered with hardboard inside, so as to convert the hut into a darkened observatory but still giving the outward appearance of a summer house.

The upper plywood ring carries the roof which consists of eight sections. Two of them are practically rectangular and measure 0.7 m (2 ft 4 in) × 1.8 m (6 ft); they are on opposite sides, one of them contain-

Figure 24.4 Side view of the centralising assembly.

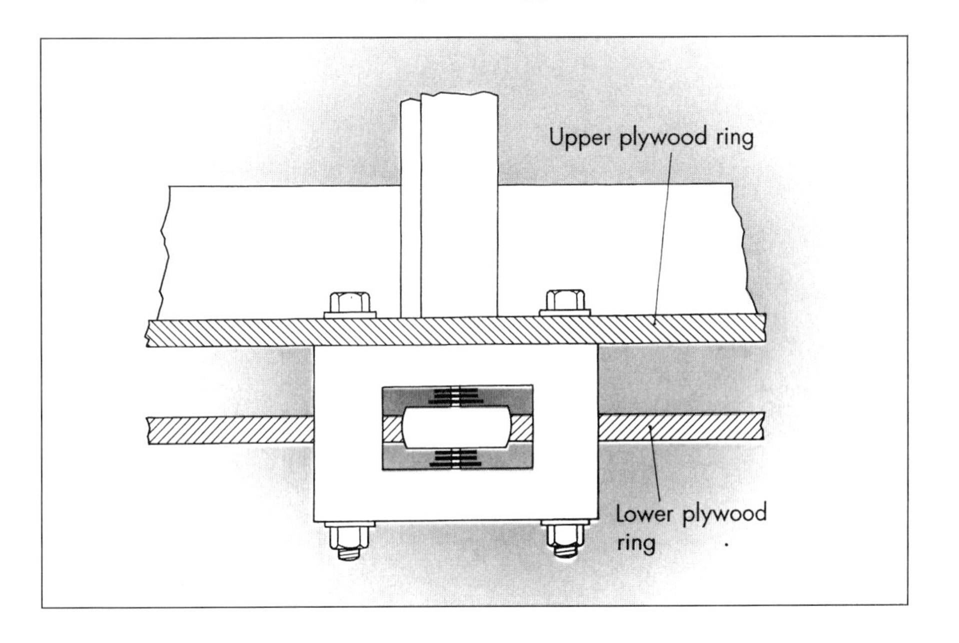

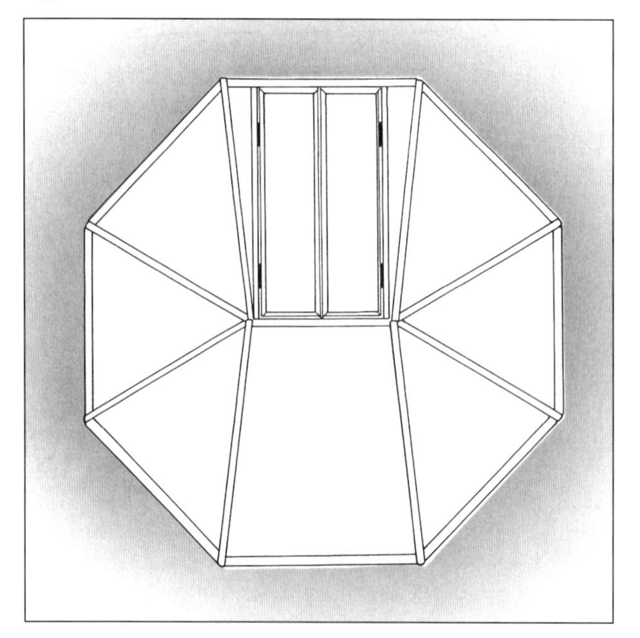

Figure 24.5 Plan view of the roof.

ing the shutter. The remaining six sections are triangular, three on either side as illustrated in Figure 24.5.

The marine ply roof was varnished and thus gave a very pleasing appearance; however the varnish tended to lift with time and after two years I had to cover the roof with roofing felt.

Uses

As a dedicated and regular solar observer, I have been using the observatory on an average of 230 days each year,. mainly for white-light observations. In 1975 I built a promscope and added this instrument to the existing mounting. In 1977 I acquired a narrow-band filter and its performance outstripped that of the promscope to the extent of superseding it and I use it with an 8 in (200 mm) Schmidt-Cassegrain telescope outside the observatory. It is stored together with many accessory items in the observatory.

Two years ago I installed a solar radio-telescope beneath the tripod.

Over the years I added many amenities to the observatory, notably electric light and mains power supply, carpets, and a telephone. In 1989 I acquired a CCD module which was converted into a CCD camera which I used to record Ha features of the Sun; as the monitor cannot be used in the brightness of the day, I have used it in the darkened solar observatory to great advantage.

The observatory serves me well for solar work but it could also be used as a general astronomical observatory.

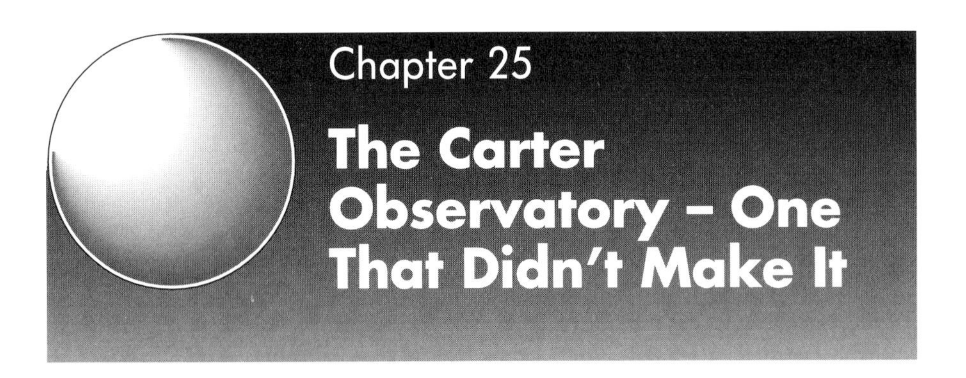

Chapter 25
The Carter Observatory – One That Didn't Make It

John Watson

Some years ago, a good friend of mine, Geoff Carter (an amateur astronomer, talented electronics engineer and one of the funniest men I ever met), set out to construct an observatory for his home-built equatorially-mounted reflector. He gave the design of his observatory a lot of thought, and eventually – he was one of nature's lateral thinkers – hit on the idea of building the walls *down* rather than up. In other words, the observatory was to be partly underground with the lower edge of the dome only just above ground level. A short flight of steps, like the steps leading to a basement flat, would provide access. The underground room would be square (and, of course, small enough to fit completely under the dome's perimeter), which would make the internal construction rather easy.

This admittedly unusual design seemed to offer a number of clear advantages. Being low in profile, the observatory building would be less obtrusive. For example, it would neither spoil his family's view of the garden nor attract as much criticism from the neighbours as a more usual observatory building would have.

Repairs to the dome – in particular work on the mechanisms for rotating it and opening and closing the shutter – would be easier and far safer, as all parts of the dome could be reached without the aid of a long ladder.

Furthermore, the subterranean design would be

inexpensive, saving the cost of foundations, bricks and a bricklayer. A fairly thin layer of concrete should suffice to keep the observer soil-free and cosy. The surrounding ground would take the considerable weight of the dome.

Having finalised his plans for the building, he set about the most difficult part of the construction – digging the hole.

This job turned out to be a lot longer and more tiring that he had expected. It took almost two weeks to complete the digging (including breaks for lunch and the occasional beer) but by the second weekend he had made a square hole about 2 m (6 ft 6 in) deep and 2.5 m (8 ft) on its sides. In the centre of this was a smaller and deeper hole to take the pier that was to support the equatorial head.

Anyone who has never seen ten cubic metres of soil and rubble might not appreciate just how much had to be shifted during the operation. The debris was eventually carted away in a large truck.

Totally exhausted, aching from head to toe, covered in blisters, but with that warm glow of a job well done, he decided to take a richly-deserved two-day fishing break. When he returned home, it was to discover that the hole had spontaneously filled up with water because most of it was below the water table.

Undaunted, he added a small waterfall and some fish.

Figure 25.1 The site of the Carter Observatory today.

Sadly, Geoff Carter died suddenly in January 1993, leaving a wife and two young daughters. His was the only funeral I have ever attended at which the floral tributes included some beautifully arranged mushrooms and some bunches of carrots. He is greatly missed.

Contributors

Dennis Allen is a computer programmer, analyst, and consultant. He is secretary of the Muskegon Astronomical Society (24 members).

John R. Fletcher FRAS is the Deep Sky Director, SPA Great Britain.

Bruce Hardie is Director of the British Astronomical Association Solar Section and observes the Sun daily in white-light and H-alpha from his observatory in Jordanstown, Co. Antrim.

M.D. (Danie) Overbeek is a member of the IAU, Past President of the Astronomical Society of Southern Africa, Director of the ASSA's Occultation Sections and Past member of Council of the American Association of Variable Star Observers. He has contributed more than 213 000 variable star observations to the AAVSO.

Lawrence and **Linda Lopez** are members of the New Hampshire Astronomical Society. He is a software engineering consultant and she is an industrial hygienist. Their email address is: lopez@mv.mv.com.

Maurice Gavin is a retired architect and British Astronomical Association President.

Alan Heath is a retired schoolteacher. He served as Director of the Saturn Section of the British Astronomical Society for thirty years. He is a regular contributor to the work of the British Astronomical Association and the Association of Lunar and Planetary Observers (USA).

Trevor Hill has been a solar astronomer for 22 years. He is Head of Science and Physics at Taunton School and Director of Taunton School Radio Astronomy Observatory.

Terry Platt is technical director of an electronics company in Maidenhead, England. He has been a member of the British Astronomical Association for about thirty years.

Chris Kitchin is the Director of Hertford University Observatory and Scheme Tutor for Physical Sciences degrees. He has written several books on telescopes, most recently *Telescopes and Techniques* in this series.

Ron Johnson is Chief Planning Engineer for a construction company. He is a member of the British Astronomical Association and the Ewell Astronomical Society.

Brian Manning is a retired engineering draughtsman. He has been a member of the British Astronomical Association for forty-eight years, and was awarded the Dall Medal for instrument making in 1990. He is one of the few modern amateurs to have discovered a new asteroid.

A.J. Sizer is Head of Junior Science at Chigwell School. He is a Fellow of the Royal Astronomical Society, a member of the British Astronomical Association and the Royal Institution. He is also a member of the Society for Popular Astronomy and is their adviser for the GCSE Astronomy examination.

David Reid is a student at Torquay Boys' Grammar School and a member of the British Astronomical Association. Since mid-1994 he has coordinated the BAA Student Group and is Editor of the BAA Student Newsletter. His main research interests are in deep sky, lunar and historical astronomy.

C. Lintott is also a student at Torquay Boys' Grammar School and an active BAA member who assists with the running of the Student Group.

Patrick Moore is an amateur astronomer who has concentrated on observations of the Moon and planets. The main telescope at his observatory in Selsey, Sussex, is a 15 in reflector. He is a Past President of the British Astronomical Association.

Jack Newton is the manager of a Marks and Spencer store in Victoria, British Columbia. He is a past president of the Winnipeg, Toronto and Victoria Centres of the Royal Astronomical Society of Canada.

David Strange is a farmer. He is a past Chairman of the Wessex Astronomical Society, UK.

Christof Plicht is a sales representative for a valve manufacturer. He is a member of the local astronomical association as well as the SIDDAS group in Sidmouth, UK.

Cheryl Power runs the Powys County Observatory with Brian Williams, for schools and the general public. It is also a venue for the Offa's Dyke Astronomical Society.

Martin Mobberley is an electronics engineer. He is also the current Papers Secretary of the British Astronomical Association and Assistant Editor of *Astronomer* magazine.

Michael Schwartz is CEO of a software company. He is a member of the local astronomical association as well as a member of the Astronomical Society of the Pacific. His email address is: pfactors@ix.netcom.com.

David Ratledge is a Civil Engineer specialising in structures and IT. He is Chairman of Bolton Astronomical Society.

E.H. Strach is a consultant orthopaedic surgeon. He is the honorary vice president of the Liverpool Astronomical Society and has been a member of the British Astronomical Association for thirty years.